RAOUL A. ROBINSON
445 Provost Lane
Fergus, Ont.   N1M 2N3

how
nature
works

COPERNICUS
AN IMPRINT OF SPRINGER-VERLAG

PER BAK

how
nature
works

the
science of
self-organized
criticality

Published in the United States by Copernicus, an imprint of
Springer-Verlag New York, Inc.

Copernicus
Springer-Verlag New York, Inc.
175 Fifth Avenue
New York, NY 10010
USA

Library of Congress Cataloging-in-Publication Data

Bak, P. (Per), 1947–
    How nature works : the science of self-organized criticality / Per Bak.
    p. cm.
    Includes bibliographical references and index.
    ISBN 0-387-94791-4 (hardcover : alk. paper)
    1. Critical phenomena (Physics)   2. Complexity (Philosophy)
3. Physics—Philosophy.   I. Title.
QC173.4.C74B34   1996
003'.7—dc20   96-16845

Manufactured in the United States of America.
Printed on acid-free paper.

9 8 7 6 5 4 3 2

ISBN 0-387-94791-4 SPIN 10560573

Who could ever calculate the path of a molecule?
How do we know that the creations of worlds are
not determined by falling grains of sand?

—Victor Hugo, *Les Miserables*

Contents

preface
and
acknowledgments

Self-organized criticality is a new way of viewing nature. The basic picture is one where nature is perpetually out of balance, but organized in a poised state—the critical state—where anything can happen within well-defined statistical laws. The aim of the science of self-organized criticality is to yield insight into the fundamental question of why nature is complex, not simple, as the laws of physics imply.

Self-organized criticality explains some ubiquitous patterns existing in nature that we view as complex. Fractal structure and catastrophic events are among those regularities. Applications range from the study of pulsars and black holes to earthquakes and the evolution of life. One intriguing consequence of the theory is that catastrophes can occur for no reason whatsoever. Mass extinctions may take place without any external triggering mechanism such as a volcanic eruption or a meteorite hitting the earth (although the theory of course cannot rule out that this has in fact occurred).

Since we first proposed the idea in 1987, more than 2,000 papers have been written on self-organized criticality, making ours the most cited paper in physics during that period. *How Nature Works* is the first book to deal with the subject. The basic idea is simple, and most of the mathematical models that have been used in the implementation of the theory are not complicated. Anyone with some computer literacy and a PC can set the models up on his own to verify the predictions. Often, no more than high school mathematics is needed. Some of the computer programs are even available on the Internet. Some of the sandpile experiments are of no greater cost and difficulty than the dedicated reader can perform him or herself. Unlike other subjects in physics, the basic ideas are simple enough to be made accessible to a non-scientific audience without being trivialized.

Many friends and colleagues have helped me, with both the research and the book. The science has been all fun—in particular I am grateful to Kurt Wiesenfeld and Chao Tang, with whom I collaborated on the original idea, and to Kan Chen, Kim Christensen, Maya Paczuski, Zeev Olami, Sergei Maslov, Michael Creutz, Michael Woodford, Dimitris Stassinopolous, and Jose Scheinkman, who participated in the research that followed, bringing the idea to life by applying it to many different phenomena in nature. Thanks are due to Elaine Wiesenfeld for drawing the logo of self-organized criticality, the sandpile, shown in Figure 1; to Ricard Sole for drawing the dog-pulling Figure 9; to Arch Johnston for providing Figure 2; to Jens Feder and his group in Oslo for Figure 6 and the figures on their ricepile experiment, Figures 15–17 and Plate 4; to Daniel Rothman and John P. Grotzinger for the photos of the Kings Peak formation, Figure 18; to Peter Grassberger for the office version of the sandpile model, Figure 13; and to Paolo Diodati for providing the original figures on the measurements of acoustic emission from Stromboli, Figure 23. The impressive computer graphics on the sandpile in Plate 1, and the "Game of Life," Plates 6–8, are due to Michael Creutz.

A number of persons helped me increase the literary qualities of the manuscript—unfortunately, the brevity of precise form that is suitable for scientific journals does not do well when addressing a broader audience. First of all, I am grateful to Maya Paczuski and Jim Niederer who spent endless hours improving the presentation and helping with organizing the material. My children, Tine and Jakob and Thomas, checked the manuscript for readability for non professionals, leading to revisions of several unclear passages. Finally, I am indebted to Jerry Lyons, William Frucht, and Robert Wexler of Copernicus Books for substantial and invaluable help with the manuscript at all stages.

complexity
and
criticality

How can the universe start with a few types of elementary particles at the big bang, and end up with life, history, economics, and literature? The question is screaming out to be answered but it is seldom even asked. Why did the big bang not form a simple gas of particles, or condense into one big crystal? We see complex phenomena around us so often that we take them for granted without looking for further explanation. In fact, until recently very little scientific effort was devoted to understanding why nature is complex.

I will argue that complex behavior in nature reflects the tendency of large systems with many components to evolve into a poised, "critical" state, way out of balance, where minor disturbances may lead to events, called avalanches, of all sizes. Most of the changes take place through catastrophic events rather than by following a smooth gradual path. The evolution to this very delicate state occurs without design from any outside agent. The state is established solely because of the dynamical interactions among individual elements of the system: the critical state is *self-organized*. Self-organized criticality is so far the only known general mechanism to generate complexity.

To make this less abstract, consider the scenario of a child at the beach letting sand trickle down to form a pile (Figure 1). In the beginning, the pile is flat, and the individual grains remain close to where they land. Their motion can be understood in terms of their physical properties. As the process continues, the pile becomes steeper, and there will be little sand slides. As time goes on, the sand slides become bigger and bigger. Eventually, some of the sand slides may even span all or most of the pile. At that point, the system is far out of balance, and its behavior can no longer be understood in terms of the behavior of the individual grains. The avalanches form a dynamic of their own, which can be understood only from a holistic description of the properties of the entire pile rather than from a reductionist description of individual grains: the sandpile is a complex system.

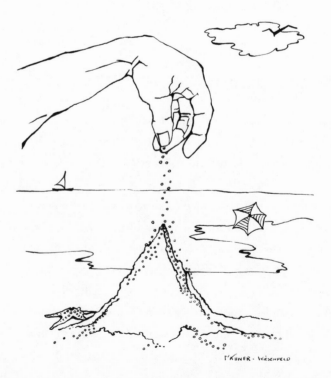

Figure 1.   Sandpile. (Drawing by Ms. Elaine Wiesenfeld.)

The complex phenomena observed everywhere indicate that nature operates at the self-organized critical state. The behavior of the critical sandpile mimics several phenomena observed across many sciences, which are associated with complexity. But before arguing that this is indeed the case, let us try to sharpen the definition of the problem. What is complexity? How have scientists and others addressed the problem in the past?

The Laws of Physics Are Simple, but Nature Is Complex

Starting from the Big Bang, the universe is supposed to have evolved according to the laws of physics. By analyzing experiments and observations, physicists have been very successful in finding those laws. The innermost secrets of matter have been revealed down to ever smaller scales. Matter consists of atoms, which are composed of elementary particles such as electrons, protons, and neutrons, which themselves are formed by quarks and gluons, and so on. All phenomena in nature, from the largest length scales spanned by the universe to the smallest represented by the quark, should be explained by the same laws of physics.

One such law is Newton's second law, $f = ma$, which simply tells us that an object that is subjected to a force responds by accelerating at a rate proportional to that force. This simple law is sufficient to describe how an apple falls to the ground, how planets orbit the sun, and how galaxies are attracted to one another by the force of gravity. Maxwell's equations describe the interactions between electrical currents and magnetic fields, allowing us to understand how an electric motor or a dynamo works. Einstein's theory of relativity says that Newton's laws have to be modified for objects moving at high velocities. Quantum mechanics tells us that electrons in an atom can only exist in states with specific energies. The electrons can jump from one state to another without spending any time in between.

These laws of physics are quite simple. They are expressed in mathematical equations that can all be written down on a couple of notebook pages. However, the mathematics involved in solving these equations, even for simple situations, may be quite complicated. This happens when there are more than two objects to consider. For instance, calculating the motion of two planets moving in the gravitational field of the other planets and the sun is exorbitantly difficult. The problem is insoluble with pen and paper, and can be done only approximately with the help of computers, but that is usually considered to be a practical problem rather than a fundamental physics problem.

The philosophy of physics since its inception has been reductionist: that the world around us can be understood in terms of the properties of simple building blocks. Even the Greeks viewed the world as consisting of only a few elements. Once we have broken the world down to its simplest fundamental laws, and the most fundamental particles have been identified, the job is complete. Once we have accomplished this feat, the role of physics—the "king of sciences"—will be played out and the stage can be left to the "lesser" sciences, such as geophysics, chemistry, and biology, to sort out the consequences.

In some special cases, physicists have succeeded in explaining the behavior of systems consisting of many parts—atoms, molecules, or electrons. For instance, the behavior of crystals, where trillions of atoms neatly occupy the rows and columns of a regular periodic lattice, is relatively well understood from the basic laws of physics. A crystal is a prime example of an "ordered" system, where each atom has its well-defined place on a regular, periodic grid. The crystal is understandable precisely because it looks the same everywhere.

At the opposite end of the spectrum from crystals are gases, which also consist of many atoms or molecules. Gases can be understood because their molecules rarely interact, by bumping into one another. In contrast to the crystal, where the atoms are ordered on a lattice, the atoms in a gas form a random, disordered system. Again, the tractability of the system arises from its uniformity. The gas looks the same everywhere, although at a given time the individual atoms at different locations move with different velocities in different directions. On average all atoms behave the same way.

However, we do not live in a simple, boring world composed only of planets orbiting other planets, regular infinite crystals, and simple gases or liquids. Our everyday situation is not that of falling apples. If we open the window, we see an entirely different picture. The surface of the earth is an intricate conglomerate of mountains, oceans, islands, rivers, volcanoes, glaciers, and earthquake faults, each of which has its own characteristic dynamics. Unlike very ordered or disordered systems, landscapes differ from place to place and from time to time. It is because of this variation that we can orient ourselves by studying the local landscape around us. I will define systems with large variability as *complex*. The variability may exist on a wide range of length scales. If we zoom in closer and closer, or look out further and further, we find variability at each level of magnification, with more and more new details appearing. In the universe, there is variability on the greatest scale. Just about every week, there is a new report from the Hubble telescope orbiting the earth, or from interplanetary satellites, on some previously undiscovered phenomenon. Complexity is a Chinese box phenomenon. In each box there are new surprises. Many different quantitative general definitions of complexity have been attempted, without much success, so let us think of complexity simply as variability. Crystals and gases and orbiting planets are not complex, but landscapes are.

As if the variability seen in astronomy and geophysics were not enough, the complexity has many more layers. Biological life has evolved on earth, with myriad different species, many with billions of individuals, competing and interacting with each other and with the environment. At the end of one tiny branch of biology we find ourselves. We can recognize other humans because we are all different. The human body and brain are formed by an intricate arrangement of interacting cells. The brain may be the most complex system of all because it can form a representation of the complex outer world. Our history, with its record of upheavals, wars, religions, and political systems, constitutes yet another level of complexity involving modern human societies with economies composed of consumers, producers, thieves, governments, and economists.

Thus, the world that we actually observe is full of all kinds of structure and surprises. How does variability emerge out of simple invariable laws? Most phenomena that we observe around us seem rather distant from the basic laws of physics. It is a futile endeavor to try to explain most natural phenomena in detail by starting from particle physics and following the trajectories of all particles. The combined power of all the computers in the world does not even come close to the capacity needed for such an undertaking.

The fact that the laws of physics specify everything (that they are deterministic) is irrelevant. The dream arising from the breathtaking progress of physics during the last two centuries combined with the advances of modern high-speed computers—that everything can be understood from "first principles"—has been thoroughly shattered. About thirty years ago, in the infancy of the computer era, there was a rather extensive effort, known as *limits to growth*, that had the goal of making global predictions. The hope was to be able to forecast, among other things, the growth of the human population and its impact on the supply of natural resources. The project failed miserably because the outcome depended on unpredictable factors not explicitly incorporated into the program. Perhaps predictions on global warming fall into the same category, since we are dealing with long-term predictions in a complex system, even though we have a good understanding of the physics of weather.

The laws of physics can explain how an apple falls but not why Newton, a part of a complex world, was watching the apple. Nor does physics have much to say about the apples origin. Ultimately, though, we believe that all the complex phenomena, including biological life, do indeed obey physical laws: we are simply unable to make the connection from atoms in which we know that the laws are correct, through the chemistry of complicated organic molecules, to the formation of cells, and to the arrangement of those cells into living organisms. There has never been any proof of a metaphysical process not following the laws of physics that would distinguish living matter from any other. One might wonder whether this state of affairs means that we cannot find general "laws of nature" describing why the ordinary things that we actually observe around us are complex rather than simple.

The question of the origin of complexity from simple laws of physics—maybe the biggest puzzle of all—has only recently emerged as an active science. One reason is that high-speed computers, which are essential in this study, have not been generally available before. However, even now the science of complexity is shrouded in a good deal of skepticism—it is not clear how any general result can possibly be helpful, because each science works well within its own domain.

Because of our inability to directly calculate how complex phenomena at one level arise from the physical mechanisms working at a deeper level, scientists sometimes throw up their hands and refer to these phenomena as "emergent." They just pop out of nowhere. Geophysics emerges from astrophysics. Chemistry emerges from physics. Biology emerges from chemistry and geophysics, and so on. Each science develops its own jargon, and works with its own objects and concepts. Geophysicists talk about tectonic plate motion and earthquakes without reference to astrophysics, biologists describe the properties and evolution of species without reference to geophysics, economists describe human monetary transactions without reference to biology, and so on. There is nothing wrong with that! Because of the seeming intractability of emergent phenomena, no other modus operandi is possible. If no new phenomena emerged in large systems out of the dynamics of systems working at a lower level, then we would need no scientists but particle physicists, since there would be no other areas to cover. But then there would be no particle physicists. Quality, in some way, emerges from quantity—but how? First let us review a couple of previous approaches to dealing with complex phenomena.

Storytelling Versus Science

The reductionist methods of physics—detailed predictions followed by comparison with reproducible experiments—are impossible in vast areas of scientific interest. The question of how to deal with this problem has been clearly formulated by the eminent paleontologist and science writer Stephen Jay Gould in his book *Wonderful Life:*

> How should scientists operate when they must try to explain the result of history, those inordinately complex events that can occur but once in detailed glory? Many large domains of nature—cosmology, geology, and evolution among them—must be studied with the tools of history. The appropriate methods focus on narrative, not experiment as usually conceived.

Gould throws up his hands and argues that only "storytelling" can be used in many sciences because particular outcomes are contingent on many single and unpredictable events. Experiments are irrelevant in evolution or paleontology, because nothing is reproducible. History, including that of evolution, is just "one damned thing after another." We can explain in hindsight what has happened, but we cannot predict what will happen in the future. The Danish philosopher Sören Kierkegaard expressed the same view in his famous phrase "Life is understood backwards, but must be lived forwards [*Livet forstaas baglaens, men maa leves forlaens*]."

Sciences have traditionally been grouped into two categories: hard sciences, in which repeatable events can be predicted from a mathematical formalism expressing the laws of nature, and soft sciences, in which, because of their inherent variability, only a narrative account of distinguishable events post mortem is possible. Physics, chemistry, and molecular biology belong to the first category; history, biological evolution, and economics belong to the second.

Gould rightfully attributes the variability of things, and therefore their complexity, to *contingency*. Historical events depend on freak accidents, so if the tape of history is replayed many times with slightly different initial conditions, the outcome will differ vastly each time. The mysterious occurrences of incidents leading to dramatic outcomes have fascinated historians and inspired fiction writers. Real life's dependence on freak events allows the fiction writer a huge amount of freedom, without losing credibility.

Historians explain events in a narrative language where event A leads to event B and C leads to D. Then, because of event D, event B leads to E. However, if the event C had not happened, then D and E would not have happened either. The course of history would have changed into another sequence of events, which would have been equally well explainable, in hindsight, with a different narrative. The discovery of America involved a long series of events, each of crucial historical importance for the actual outcome: Columbus' parents had to meet each other, Columbus had to be born, he had to go to Spain to get funding, the weather had to be reasonable, and so on. History is unpredictable, but not unexplainable. There is nothing wrong with this way of doing science, in which the goal is an accurate narrative account of specific events. It is precisely the overwhelming impact of contingency that makes those sciences interesting. There will always be more surprises in store for us. In contrast, simple predictable systems, such as an apple falling to the ground, become boring after a while.

In the soft sciences, where contingency is pervasive, detailed long-term prediction becomes impossible. A science of evolutionary biology, for example, cannot explain why there are humans and elephants. Life as we see it today is just one very unlikely outcome among myriad other equally unlikely possibilities. For example, life on earth would be totally different if the dinosaurs had not become extinct, perhaps as a consequence of a meteor hitting the earth instead of continuing in its benign orbit. An unlikely event is likely to happen because there are so many unlikely events that could happen.

But what underlying properties of history and biology make them sensitive to minor accidental events? In other words, what is the underlying nature of the dynamics that leads to the interdependence of events and thus to complexity? Why can incidents happen that have dramatic global consequences? Why the dichotomy of the sciences into two quite disparate groups with different methods and styles, since presumably all systems in the final analysis obey the same laws of nature?

Before going into the details of the theory, let us explore, in general terms, what a science of complexity could be.

What Can a Theory of Complexity Explain?

If all that we can do in the soft, complex sciences is to monitor events and make short-term predictions by massive computations, then the soft sciences are no place for physicists to be, and they should gracefully leave the stage for the "experts" who have detailed knowledge about their particular fields. If one cannot predict anything specific, then what is the point?

In a well-publicized debate in January 1995 at the Linnean institute in London, between the biologist Stuart Kauffman of the Santa Fe Institute, and John Maynard Smith of the University of Sussex, England, author of *The Theory of Evolution*, Smith exclaimed that he did not find the subject of complexity interesting, precisely because it has not explained any detailed fact in nature.

Indeed, any theory of complexity must necessarily appear insufficient. The variability precludes the possibility that all detailed observations can be condensed into a small number of mathematical equations, similar to the fundamental laws of physics. At most, the theory can explain *why* there is variability, or what typical patterns may emerge, not what the particular outcome of a particular system will be. The theory will never predict elephants. Even under the most optimistic circumstances, there will still be room for historians and fiction writers in the future.

A general theory of complex systems must necessarily be *abstract*. For example, a theory of life, in principle, must be able to describe all possible scenarios for evolution. It should be able to describe the mechanisms of life on Mars, if life were to occur. This is an extremely precarious step. Any general model we might construct cannot have any specific reference to actual species. The model may, perhaps, not even refer to basic chemical processes, or to the DNA molecules that are integral parts of any life form that we know.

We must learn to free ourselves from seeing things the way they are! A radical scientific view, indeed! If, following traditional scientific methods, we concentrate on an accurate description of the details, we lose perspective. A theory of life is likely to be a theory of a process, not a detailed account of utterly accidental details of that process, such as the emergence of humans.

The theory must be *statistical* and therefore cannot produce specific details. Much of evolutionary theory, as presented for instance in Maynard Smith's book, is formulated in terms of anecdotal evidence for the various mechanisms at work. Anecdotal evidence carries weight only if enough of it can be gathered to form a statistical statement. Collecting anecdotal evidence can only be an intermediate goal. In medicine, it was long ago realized that anecdotal evidence from a single doctor's observation must yield to evidence based on a large, statistically significant set of observations. Confrontation between theories and experiments or observations, essential for any scientific endeavor, takes place by comparing the statistical features of general patterns.

The *abstractness* and the *statistical, probabilistic nature* of any such theory might appear revolting to geophysicists, biologists, and economists, expecting to aim for photographic characterization of real phenomena.

Perhaps too much emphasis has been put on *detailed* prediction, or forecasting, in science in today's materialistic world. In geophysics, the emphasis is on predicting specific earthquakes or other disasters. Funding is provided according to the extent to which the budget agencies and reviewers judge that progress might be achieved. This leads to charlatanism and even fraud, not to mention that good scientists are robbed of their grants. Similarly, the emphasis in economics is on prediction of stock prices and other economic indicators, since accurate predictions allow you to make money. Not much effort has been devoted to describing economic systems in an unbiased, detached way, as one would describe, say, an ant's nest.

Actually, physicists are accustomed to dealing with probabilistic theories, in which the specific outcome of an experiment cannot be predicted—only certain statistical features. Three fundamental theories in physics are of a statistical nature. First, statistical mechanics deals with large systems in equilibrium, such as the gas of atoms in the air surrounding us. Statistical mechanics tells us how to calculate average properties of the many atoms forming the gas, such as the temperature and the pressure. The theory does not give us the positions and the velocities of all the individual atoms (and we couldn't care less anyhow). Second, quantum mechanics tells us that we cannot predict both the specific position and velocity of a small particle such as an electron at the same time, but only the probability that an experiment would find the particle at a certain position. Again, we are most often interested only in some average property of many electrons, as for instance the electric current through a wire, which may again be predictable. Third, chaos theory tells us that many simple mechanical systems, for example pendulums that are pushed periodically, may show unpredictable behavior. We don't know exactly where the pendulum will be after a long time, no matter how well we know the equations for its motion and its initial state.

As pointed out by the philosopher Karl Popper, prediction is our best means of distinguishing science from pseudoscience. To predict the statistics of actual phenomena rather than the specific outcome is a quite legitimate and ordinary way of confronting theory with observations.

What makes the situation for biology, economics, or geophysics conceptually different, and what makes it more difficult to accept this state of affairs, is that the outcome of the process is *important*. As humans, we care about the specific state of the system. We don't just observe the average properties of many small unpredictable events, but only one specific outcome in its full glory. The fact that we may understand the statistical properties of earthquakes, such as the average number of earthquakes per year of a certain size in a certain area, is of little consolation to those who have been affected by large,

devastating earthquakes. In biology, it is important that the dinosaur vanished during a large extinction event and made room for us.

Psychologically, we tend to view our particular situation as unique. It is
emotionally unacceptable to view our entire existence as one possible fragile
outcome among zillions of others. The idea of many parallel possible universes is hard to accept, although it has been used by several science-fiction
writers. The problem with understanding our world is that we have nothing
to compare it with.

We cannot overcome the problem of unpredictability. Kierkegaard's philosophy represents the fundamental and universal situation of life on earth.
So how can there be a general theory or science of complexity? If such a theory
cannot explain any specific details, what is the theory supposed to explain?
How, precisely, can one confront theory with reality? Without this crucial
step, there can be no science.

Fortunately, there are a number of ubiquitous general empirical observations across the individual sciences that cannot be understood within the set
of references developed within the specific scientific domains. These phenomena are the occurrence of large catastrophic events, fractals, one-over-f noise
($1/f$ noise), and Zipf's law. A litmus test of a theory of complexity is its ability
to explain these general observations. Why are they universal, that is, why do
they pop up everywhere?

Catastrophes Follow a Simple Pattern

Because of their composite nature, complex systems can exhibit catastrophic
behavior, where one part of the system can affect many others by a domino
effect. Cracks in the crust of the earth propagate in this way to produce earthquakes, sometimes with tremendous energies.

Scientists studying earthquakes look for specific mechanisms for large
events, using a narrative individual description for each event in isolation
from the others. This occurs even though the number of earthquakes of a
given magnitude follows a glaringly simple distribution function known as
the Gutenberg–Richter law. It turns out that every time there are about 1,000
earthquakes of, say, magnitude 4 on the Richter scale, there are 100 earthquakes of magnitude 5, 10 of magnitude 6, and so on. This law is illustrated in

Figure 2a, which shows how many earthquakes there were of each magnitude in a region of the southeastern United States known as the New Madrid earthquake zone during the period 1974–1983. Figure 2b shows where those earthquakes took place. The size of the dots represents the magnitudes of the earthquakes. The information contained in the figures was collected by Arch C. Johnston and Susan Nava of the Memphis State University. The scale is a logarithmic one, in which the numbers on the vertical axis are 10, 100, 1,000 instead of 1, 2, 3. The Gutenberg-Richter law manifests itself as a straight line in this plot.

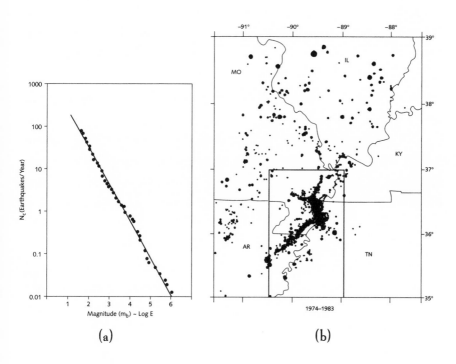

(a)   (b)

Figure 2.   (a) Distribution of earthquake magnitudes in the New Madrid zone in the southeastern United States during the period 1974–1983, collected by Arch Johnston and Susan Nava of Memphis State University. The points show the number of earthquakes with magnitude larger than a given magnitude *m*. The straight line indicates a power law distribution of earthquakes. This simple law is known as the Gutenberg–Richter law. (b) Locations of the earthquakes used in the plot. The size of the dots represent the magnitudes of the earthquakes.

The horizontal x-axis is also logarithmic, since the magnitude m measures the logarithm of the energy released by the earthquake, rather than the energy itself. Thus, an earthquake of magnitude 6 is ten times stronger than an earthquake of magnitude 5, and an earthquake of magnitude 4 is ten times stronger than an earthquake of magnitude 3. An earthquake of magnitude 8 is 10 million times more energetic than one of magnitude 1, which corresponds to a large truck passing by. By using worldwide earthquake catalogues, the straight line can be extended to earthquakes of magnitudes 7, 8, and 9. This law is amazing! How can the dynamics of all the elements of a system as complicated as the crust of the earth, with mountains, valleys, lakes, and geological structures of enormous diversity, conspire, as if by magic, to produce a law with such extreme simplicity? The law shows that large earthquakes do not play a special role; they follow the same law as small earthquakes. Thus, it appears that one should not try to come up with specific explanations for large earthquakes, but rather with a general theory encompassing all earthquakes, large and small.

The importance of the Gutenberg–Richter law cannot be exaggerated. *It is precisely the observation of such simple empirical laws in nature that motivates us to search for a theory of complexity.* Such a theory would complement the efforts of geophysicists who have been occupied with their detailed observations and theorizing on specific large earthquakes and fault zones without concern about the general picture. One explanation for each earthquake, or for each fault.

In their fascinating book *Tales of the Earth,* Officer and Page argue that the regularity of numerous catastrophic phenomena on earth, including flooding, earthquakes, and volcanic eruptions, has a message for us on the basic mechanisms driving the earth, which we must unravel in order to deal with those phenomena (or, perhaps, to understand why we cannot deal with them).

In economics, an empirical pattern similar to the Gutenberg–Richter law holds. Benoit Mandelbrot, of IBM's T. J. Watson Center in New York, pointed out in 1966 that the probability of having small and large variations on prices of stocks, cotton, and other commodities follows a very simple pattern, known as a Levy distribution. Mandelbrot had collected data for the variation of cotton prices from month to month over several years. He then counted how often the monthly variation was between 10 and 20 percent, how often the variation was between 5 and 10 percent, and so on, and plotted the results on a logarithmic plot (Figure 3). Just as Johnston and Nava counted how many earthquakes there were of each size, Mandelbrot counted how many months there were with a given price variation. Note the smooth transition from small variations to large ones. The distribution of price changes follows approximately a straight line, a power law. The price variations are "scale free" with no typical size of the variations, just as earthquakes do not have a typical characteristic size.

Mandelbrot studied several different commodities, and found that they all followed a similar pattern, but he did not speculate about the origin of the regular behavior that he observed. Economists have chosen largely to ignore Mandelbrot's work, mostly because it doesn't fit into the generally accepted picture. They would discard large events, since these events can be attributed to specific "abnormal circumstances," such as program trading for the crash of October 1987, and excessive borrowing for the crash of 1929. Contingency

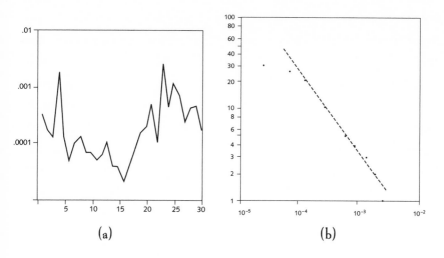

(a)   (b)

Figure 3.   (a) Monthly variations of cotton prices (Mandelbrot, 1963) during a period of 30 months. (b) The curve shows the number of months where the relative variation exceeded a given fraction. Note the smooth transition from small variations to large variations. The straight line indicates a power law. Other commodities follow a similar pattern.

is used as an argument for statistical exclusion. Economists often "cull" or "prune" the data before analysis. How can there be a general theory of events that occur once? However, the fact that large events follow the same law as small events indicates that there is nothing special about those events, despite their possibly devastating consequences.

Similarly, in biological evolution, Professor David Raup of the University of Chicago has pointed out that the distribution of extinction events follows a smooth distribution where large events, such as the Cretaceous extinction of dinosaurs and many other species, occur with fairly well defined probability and regularity. He used data collected by Jack Sepkoski, who had spent "ten years in the library" researching the fossil records of thousands of marine species. Sepkoski split geological history into 150 consecutive periods of 4 million years. For each period, he estimated what fraction of species had disappeared since the previous period (Figure 4). The estimate is a measure of the extinction rate. Sometimes there were very few extinctions, less than 5 percent,

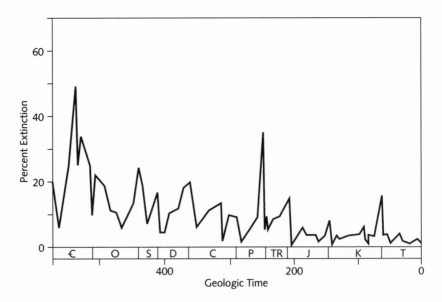

Figure 4. Biological extinctions over the last 600 million years as recorded by John Sepkoski, Jr. who spent 10 years in the library collecting the data from the fossil record. The curve shows the estimated percentage of families that went extinct within intervals of approximately 4 million years (Sepkoski, 1993).

and sometimes there were more than 50 percent extinctions. The famous Cretaceous event in which the dinosaurs became extinct is not even among the most prominent. Raup simply counted the number of periods in which the relative number of extinctions was less than 10 percent, how many periods the variation was between 10 and 20 percent, and so on, and made a histogram (Figure 5). This is the same type of analysis that Mandelbrot made for cotton prices: extinction rates replace price variations, 4-million-year intervals replace

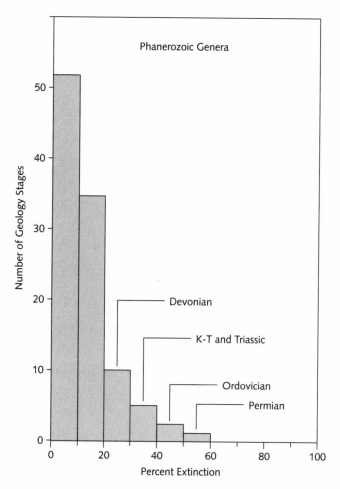

Figure 5.   Histogram of the extinction events from Figure 4 as shown by Raup. The diagram shows the number of four-million-year periods where the extinction rate was within a given range. The large well-known extinction events appear in the tail of the curve.

monthly ones. The resulting histogram forms a smooth curve, with the number of large events extending smoothly from the much larger number of small events.

Although large events occur with a well-defined probability, this does not mean the phenomenon is periodic, as Raup thought it was. The fact that an earthquake has not taken place for a long time does not mean that one is due. The situation is similar to that of a gambling roulette. Even if on average black comes out every second time, that does not mean that the outcome alternates between black and red. After seven consecutive reds, the probability that the next event is black is still $1/2$. The same goes for earthquakes. That events occur at some average interval does not mean that they are cyclical. For example, the fact that wars happen on average, say, every thirty years, cannot be used to predict the next war. The variations of this interval are large.

Again, specific narratives may explain each large catastrophe, but the regularity, not to be confused with periodicity, suggests that the same mechanisms work on all scales, from the extinctions taking place every day, to the largest one, the Cambrian explosion, causing the extinction of up to 95 percent of all species, and, fortunately, the creation of a sufficiently compensating number of species.

That catastrophes occur at all is quite amazing. They stand in sharp contrast to the theory of uniformitarianism, or gradualism, which was formed in the last century by the geophysicist Charles Lyell in his book *Principles of Geology*. According to his theory, all change is caused by processes that we currently observe, which have worked at the same rate at all times. For instance, Lyell proposed that landscapes are formed by gradual processes, rather than catastrophes like Noah's flood, and the features that we see today were made by slow persistent processes, with time as the "great enabler" that eventually makes large changes.

Lyell's uniformitarian view appears perfectly logical. The laws of physics are generally expressed as smooth, continuous equations. Since these laws should describe everything, it is natural to expect that the phenomena that we observe should also vary in a smooth and gradual manner. An opposing philosophy, catastrophism, claims that changes take place mostly through sudden cataclysmic events. Since catastrophism smacks of creationism, it has been largely rejected by the scientific community, despite the fact that catastrophes actually take place.

Fractal Geometry

Mandelbrot has coined the word *fractal* for geometrical structures with features of all length scales, and was among the first to make the astounding observation that nature is generally fractal. Figure 6a shows the coast of Norway, which appears as a hierarchical structure of fjords, and fjords within fjords, and fjords within fjords of fjords. The question "How long is a typical fjord?" has no answer—the phenomenon is "scale free." If you see a picture of part of

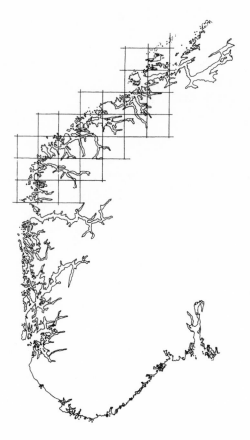

Figure 6.   (a) The coast of Norway. Note the "fractal," hierarchical geometry, with fjords, and fjords within fjords, and so on. Mandelbrot has pointed out that landscapes often are fractals. (From Feder, 1988.)

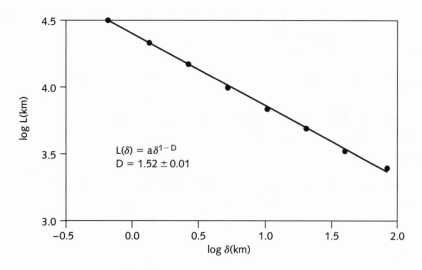

Figure 6. Continued (b) The length L of the coast measured by covering the coast with boxes, like the ones shown in (a), of various lengths δ. The straight line indicates that the coast is fractal. The slope of the line yields the "fractal dimension" of the coast of Norway, $D = 1.52$.

the fjord, or part of the coastline, you wouldn't know how large it is if the picture does not also show a ruler. Also, the length measured depends on the resolution of the ruler used for the measurement. A very large ruler that measures features only on the scale of miles will yield a much smaller estimate of the length than if a fine ruler, which can follow details on the scale of meters, is used.

One way of representing this is to measure how many boxes of a certain size δ are needed to cover the coast. Obviously, the smaller the box, the more boxes are needed to cover the coast. Figure 6b shows the logarithm of the length L measured with boxes of size δ. Had the coast been a straight line, of dimension 1, the number of boxes would be inversely proportional to δ, so the measured length would be independent of δ, and the curve would be flat. If you measure the length of a line, it doesn't matter what the size of the ruler is. However, the number of boxes needed grows much faster than that since the boxes have to follow the wrinkles of the coastline, so the straight line has a slope. The negative slope of the line gives the "fractal dimension" of the coast. Fractals in general have dimensions that are not simple integer numbers. Here, one finds $D = 1.52$, showing that the coast is somewhere between a straight line with dimension 1 and a surface of dimension 2.

A mountain range includes peaks that may range from centimeters to kilometers. No size of mountain is typical. Similarly, there are clouds of all sizes, with large clouds looking much like enlarged versions of small clouds. The universe consists of galaxies, and clusters of galaxies, and clusters of clusters of galaxies, and so on. No size of fjord, mountain, or cloud is the "right" size.

A lot of work has been done characterizing the geometrical properties of fractals, but the problem of the dynamical origin of fractals persists—where do they come from? "Fractals: Where is the Physics?" Leo Kadanoff of the University of Chicago asked in a famous editorial in *Physics Today* in 1987. Unfortunately, the title was generally viewed as a rhetorical dismissal of the whole concept of fractals rather than a legitimate cry for an understanding of the phenomenon.

The importance of Mandelbrot's work parallels that of Galileo, who observed that planets orbit the sun. Just as Newton's laws are needed to explain planetary motion, a general theoretical framework is needed to explain the fractal structure of Nature. Nothing in the previously known general laws of physics hints at the emergence of fractals.

"One-Over-f" Noise: Fractals in Time

A phenomenon called $1/f$ (one-over-f) "noise" has been observed in systems as diverse as the flow of the river Nile, light from quasars (which are large, faraway objects in the universe), and highway traffic. Figure 7a shows the light from a quasar measured over a period of eighty years. There are features of all sizes: rapid variations over minutes, and slow variations over years. In fact, there seems to be a gradual decrease over the entire period of eighty years, which might lead to the erroneous identification of a general tendency toward decreasing intensity within a human lifetime, a tendency that needs explanation.

The signal can be seen as a superposition of bumps of all sizes; it looks like a mountain landscape in time, rather than space. The signal can, equivalently, be seen as a superposition of periodic signals of all frequencies. This is another way of stating that there are features at all time scales. Just as Norway has fjords of all sizes, a $1/f$ signal has bumps of all durations. The strength or "power" of its frequency component is larger for the small frequencies; it is inversely proportional to the frequency, f. That is why we call it $1/f$ noise, although it might be misleading to call it noise rather than signal. A simple example is the velocity of a car driving along a heavily trafficked highway. There are periods of stop and go of all lengths of time, corresponding to traffic jams of all sizes. The British geophysicist J. Hurst spent a lifetime studying the water level of the Nile. Again, the signal is $1/f$, with intervals of high levels extending over short, intermediate, and long periods.

Figure 7 also shows the record of global average temperature variation on earth over the same period. This record is rising over roughly the same period as the quasar intensity decreases. One could conclude that the changes of quasar intensity and global temperature are correlated, but most reasonable people would not. In fact, the temperature variations can also be interpreted as $1/f$ noise. The apparent increase in temperature might well be a statistical fluctuation rather than an indication of global warming generated by human activity. Amusingly, Dr. Richard Voss of IBM has demonstrated that the variations in music have a $1/f$ spectrum. Maybe we write music to mirror nature.

One-over-f noise is different from random white noise, in which there are no correlations between the value of the signal from one moment to the next. In Figure 7c the white noise pattern has no slow fluctuations, that is, no large bumps. White noise sounds like the hiss on the radio in between stations rather than music, and includes all frequencies in an equal amount. A simple periodic behavior with just one frequency would be just one tone continuing forever. The $1/f$ noise lies between these two extremes; it is interesting and complex, whereas white noise is simple and boring. Amazingly, despite the fact that $1/f$ noise is ubiquitous, there has been no general understanding of its origin. It has been one of the most stubborn problems in physics. Sometimes the spectrum is not $1/f$, but $1/f^{\alpha}$, where α is an exponent with a value between 0 and 2. Such spectra are also commonly referred to as $1/f$ noise.

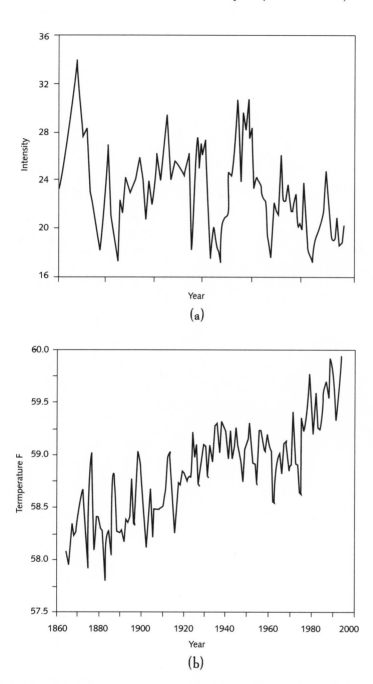

Figure 7. (a) Light emitted from a quasar during a period of 80 years from 1887–1967 (Press, 1978). Note the pattern of fast, slow, and intermediate range fluctuations. This type of signal is known as one-over-f noise ($1/f$ noise), and is extremely common in nature. (b) Global temperature monitored since 1865 (NASA).

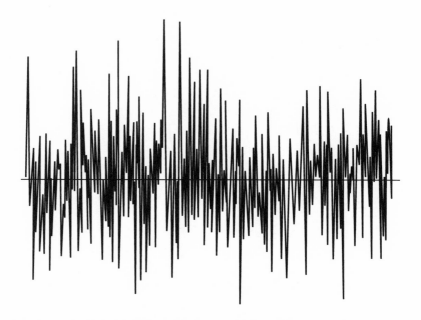

Figure 7. Continued (c) For comparison, a "boring" random, white noise pattern is also shown. This pattern has no slow fluctuations, i.e., no large bumps.

Zipf's Law

In a remarkable book that came out in 1949, *Human Behavior and the Principle of Least Effort,* Professor George Kingsley Zipf of Harvard University made a number of striking observations of some simple regularities in systems of human origin. Figure 8a shows how many cities in the world (circa 1920) had more than a given number of inhabitants. There were a couple of cities larger than 8 million, ten larger than 1 million, and 100 larger than 200,000. The curve is roughly a straight line on a logarithmic plot. Note the similarity with the Gutenberg-Richter law, although, of course, the phenomena being described couldn't be more different. Zipf made similar plots for many geographical areas and found the same behavior.

Zipf also counted how often a given word was used in a piece of literature, such as James Joyce's *Ulysses* or a collection of American newspapers. The tenth most frequently used word (the word of "rank" 10) appeared 2,653 times. The twentieth most used word appeared 1,311 times. The 20,000th most frequent word was used only once. Figure 8b shows the frequency of words used in the English language versus their ranking. The word of rank 1, *the*, is used with a frequency of 9 percent. The word of rank 10, *I*, has a frequency of 1 percent, the word of rank 100, *say*, is used with a frequency of 0.1 percent, and so on. Again, a remarkable straight line emerges. It does not matter whether the data are taken from newspapers, the Bible, or *Ulysses*—the curve is the same. The regularity expressed by the straight lines in the logarithmic plot of rank versus frequency, with slope near unity, is referred to as Zipf's law.

Although Zipf does allude to the source of this regularity being the individual agent trying to minimize his effort, he gave no hints as to how to get

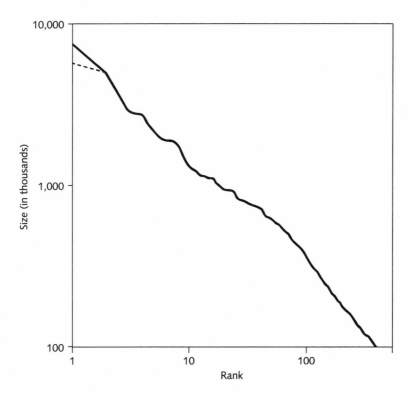

Figure 8.   (a) Ranking of cities by size around the year 1920 (Zipf, 1949). The curve shows the number of cities in which the population exceeds a given size or, equivalently, the relative ranking of cities versus their population.

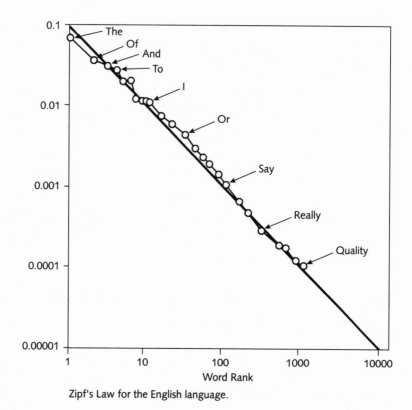

Zipf's Law for the English language.

Figure 8.   Continued (b) Ranking of words in the English language. The curve shows how many words appear with more than a given frequency.

from the individual level to the statistical observations. Zipf's law as well as the other three phenomena are emergent in the sense that they are not obvious consequences of the underlying dynamical rules.

Note that all the observations are of statistical nature. The Gutenberg-Richter law is a statement about how many earthquakes there are of each size—not where and when a particular earthquake will or did take place. Zipf's law deals with the number of cities within a given range of populations—not with why a particular city has a certain number of inhabitants. The various laws are expressed as distribution functions for measurable quantities. Therefore, a theory explaining those phenomena must also be statistical, as we have already argued.

Power Laws and Criticality

What does it mean that something is a straight line on a double logarithmic plot? Mathematically, such straight lines are called "power laws," since they show that some quantity N can be expressed as some power of another quantity s:

$$N(s) = s^{-\tau}.$$

Here, s could be the energy released by an earthquake, and $N(s)$ could be the number of earthquakes with that energy. The quantity s could equally well be the length of a fjord, and $N(s)$ could be the number of fjords of that length. Fractals are characterized by power law distributions. Taking the logarithm on both sides of the equation above we find

$$\log N(s) = -\tau \log s.$$

This shows that $\log N(s)$ plotted versus $\log s$ is a straight line. The exponent τ is the slope of the straight line. For instance, in Zipf's law the number N of cities with more than s inhabitants was expressed as $N(s) = 1/s = s^{-1}$. That is a power law with exponent -1. Essentially all the phenomena to be discussed in this book can be expressed in terms of power laws. The scale invariance can be seen from the simple fact that the straight line looks the same everywhere. There are no features at some scale that makes that particular scale stand out. There are no kinks or bumps anywhere. Of course, this must eventually break down at small and large scales. There are no fjords larger than Norway, and no fjords smaller than a molecule of water. But in between these two extremes there are features of all scales. In his beautiful book *Fractals, Chaos, Power Laws: Minutes from an Infinite Paradise,* Manfred Schroeder reviews the abundance and significance of power laws in nature.

Thus, the problem of explaining the observed statistical features of complex systems can be phrased mathematically as the problem of explaining the underlying power laws, and more specifically the values of the exponents. Let us first, however, consider a couple of approaches that have proven unsuccessful.

Systems in Balance Are Not Complex

Physicists have had some experience in dealing with large "many body" systems, in particular with systems that are in balance in a stable equilibrium. A gas of atoms and the sand at a flat beach are large systems in equilibrium; they are "in balance." If an equilibrium system is disturbed slightly, for instance by pushing a grain of sand somewhere, not much happens. In general, *systems in balance do not exhibit any of the interesting behavior discussed above, such as large catastrophes, 1/f noise, and fractals.*

There is one minor reservation. A closed equilibrium system can show complex behavior characterized by power laws, but only under very special circumstances. There has been spectacular progress in our understanding of systems at a phase transition where the system goes from a disordered state to an ordered state, for instance when the temperature is varied. Right at the critical point separating these two phases there is complex behavior characterized by scale-free behavior, with ordered domains of all sizes. To reach the critical point, the temperature has to be tuned very accurately in order to have complex behavior. But outside the laboratory no one is around to tune the parameter to the very special critical point, so this does not provide insight into the widespread occurrence of complexity in nature.

In the past, it has often been more or less tacitly assumed that large systems, such as those we find in biology and economics, are in a stable balance, like the sand at a flat beach. The leading economic theory up to now, the general equilibrium theory, assumes that perfect markets, perfect rationality, and so on bring economic systems into stable Nash equilibria in which no agent can improve his situation by any action. In the equilibrium state, small perturbations or shocks will cause only small disturbances, modifying the equilibrium state only slightly. The system's response is proportional to the size of the impact; equilibrium systems are said to be "linear." *Contingency* is irrelevant. Small freak events can never have dramatic consequences. Large fluctuations in equilibrium systems can occur only if many random events accidentally pull in the same direction, which is prohibitively unlikely. Therefore, equilibrium theory does not explain much of what is actually going on, such as why stock prices fluctuate the way they do.

A general equilibrium theory has not been explicitly formulated for biology, but a picture of nature as being in "balance" often prevails. Nature is supposed to be something that can, in principle, be conserved; this idea motivates environmentalists and conservationists. No wonder: in a human lifetime the natural world changes very little, so equilibrium concepts may seem natural or intuitive. However, if nature is in balance, how did we get here in the first place? How can there be evolution if things are in balance? Systems in balance or equilibrium, by definition, do not go anywhere. Does nature as we see it now (or a few years ago before we "started" polluting our environment) have any preferential status from an evolutionary point of view? Implicitly, the idea of nature being in balance is intimately related to the view that humans are at the center: our natural world is the "right one."

As pointed out by Gould and Eldridge, the apparent equilibrium is only a period of tranquillity, or stasis, between intermittent bursts of activity and volatility in which many species become extinct and new ones emerge. Also, the rate of evolution of individual species, as measured, for instance, by their change in size, takes place episodically in spurts. This phenomenon is called *punctuated equilibrium*. The concept of punctuated equilibrium turns out to be at the heart of the dynamics of complex systems. Large intermittent bursts have no place in equilibrium systems, but are ubiquitous in history, biology, and economics.

None of the phenomena described above can be explained within an equilibrium picture. On the other hand, no general theory for large nonequilibrium systems exists. The legendary Hungarian mathematician John von Neumann once referred to the theory of nonequilibrium systems as the "theory of non-elephants," that is, there can be no unique theory of this vast area of science.

Nevertheless, such a theory of non-elephants will be attempted here. The picture that we should keep in mind is that of a steep sandpile, emitting avalanches of all sizes, contrasting with the equilibrium flat sand box.

Chaos Is Not Complexity

In the 1980s a revolution occurred in our understanding of simple dynamical systems. It had been realized for some time that systems with a few degrees of freedom could exhibit chaotic behavior. Their future behavior is unpredictable no matter how accurately one knows their initial state, even if we had perfect knowledge of the equations that govern their motion, as we have for the swing, or a pendulum, being pushed at regular intervals.

The revolution was triggered by Mitch Feigenbaum of Los Alamos National Laboratory, a scientist working in an environment similar to mine. He had constructed a simple and elegant theory for the transition to chaos for a simple model of a predator-prey system. The model was actually invented several years earlier by the British biologist Robert May. The number of individuals, x_n, who are alive one year can be related to the number of species that are alive the following year, x_{n+1}, by a simple "map":

$$x_{n+1} = \lambda x_n(1 - x_n).$$

Feigenbaum studied this map using a simple pocket calculator. Starting with a random value of x_n, the map was used repeatedly to generate the populations at subsequent years. For small values of the parameter λ, the procedure would eventually approach a fixed point at which the population remains constant ever after. For larger values the map goes into a cycle in which every second year the population returns to the same value. For even larger values of λ the map first goes into a four-cycle, then an eight-cycle, until at some point (the Feigenbaum point) it goes into a completely chaotic state. In the chaotic phase, a small uncertainty in the initial value of the population is amplified as time passes, precluding predictability. Feigenbaum constructed a beautiful mathematical theory of this scenario. This was the first theory of the transition from regular periodic behavior to chaos. Chaos theory shows how simple systems can have unpredictable behavior.

Chaos signals have a white noise spectrum, not $1/f$. One could say that chaotic systems are nothing but sophisticated random noise generators. If the value of x (or the position of the regularly pushed swing) is plotted versus time, the signal looks much like the noise shown in Figure 7c. It is random and boring. Chaotic systems have no memory of the past and cannot evolve. However, precisely at the "critical" point where the transition to chaos occurs, there is complex behavior, with a $1/f$-like signal (Figure 7a). The complex state is at the border between predictable periodic behavior and unpredictable chaos. Complexity occurs only at one very special point, and not for the general values of λ where there is real chaos. The complexity is not robust! Since all the empirical phenomena we have discussed—fractals, $1/f$ noise, catastrophes, and Zipf's law—occur ubiquitously, they cannot depend on some delicate selection of temperature, pressure, or whatever, as represented by the parameter λ. Borrowing a metaphor from Dawkins, who got it from the English theologian William Palay, nature is operated by a "blind watchmaker" who is unable to make continuous fine adjustments.

Also, simple chaotic systems cannot produce a spatial fractal structure like the coast of Norway. In the popular literature, one finds the subjects of chaos and fractal geometry linked together again and again, despite the fact that they have little to do with each other. The confusion arises from the fact that chaotic motion can be described in terms of mathematical objects known as *strange attractors* embedded in an abstract phase space. These strange attractors have fractal properties, but they do not represent geometrical fractals in real space like those we see in nature.

In short, chaos theory cannot explain complexity.

Self-Organized Criticality

The four phenomena discussed here—regularity of catastrophic events, fractals, $1/f$ noise, and Zipf's law—are so similar, in that they can all be expressed as straight lines on a double logarithmic plot, that they make us wonder if they are all manifestations of a single principle. Can there be a Newton's law, $f = ma$, of complex behavior? Maybe self-organized criticality is that single underlying principle.

Self-organized critical systems evolve to the complex critical state without interference from any outside agent. The process of self-organization takes place over a very long transient period. Complex behavior, whether in geophysics or biology, is always created by a long process of evolution. It cannot be understood by studying the systems within a time frame that is short compared with this evolutionary process. The phrase "you cannot understand the present without understanding history" takes on a deeper and more precise meaning. The laws for earthquakes cannot be understood just by studying earthquakes occurring in a human lifetime, but must take into account geophysical processes that occurred over hundreds of millions of years and set the stage for the phenomena that we are observing. Biological evolution cannot be understood by studying in the laboratory how a couple of generations of rats or bacteria evolve.

The canonical example of SOC is a pile of sand. A sandpile exhibits punctuated equilibrium behavior, where periods of stasis are interrupted by intermittent sand slides. The sand slides, or avalanches, are caused by a domino effect, in which a single grain of sand pushes one or more other grains and causes them to topple. In turn, those grains of sand may interact with other grains in a chain reaction. Large avalanches, not gradual change, make the link between quantitative and qualitative behavior, and form the basis for emergent phenomena.

If this picture is correct for the real world, then we must accept instability and catastrophes as inevitable in biology, history, and economics. Because the outcome is contingent upon specific minor events in the past, we must also abandon any idea of detailed long-term determinism or predictability. In economics, the best we can do, from a selfish point of view, is to shift disasters to our neighbors. Large catastrophic events occur as a consequence of the same dynamics that produces small ordinary everyday events. This observation runs counter to the usual way of thinking about large events, which, as we have seen, looks for specific reasons (for instance, a falling meteorite causing the extinction of dinosaurs) to explain large cataclysmic events. Even though there are many more small events than large ones, most of the changes of the system are associated with the large, catastrophic events. Self-organized criticality can be viewed as the theoretical justification for catastrophism.

the discovery
of self-organized
criticality

In 1987 Chao Tang, Kurt Wiesenfeld, and I constructed the simple, proto-typical model of self-organized criticality, the sandpile model. Our calculations on the model showed how a system that obeys simple, benign local rules can organize itself into a poised state that evolves in terms of flashing, intermittent bursts rather than following a smooth path. We did not set out with the intention of studying sandpiles. As with many other discoveries in science, the discovery of sandpile dynamics was accidental. This chapter describes the events leading to the discovery. In hindsight, things could have been much simpler; our thinking went through some quite convoluted paths.

Science at Brookhaven

We were working at Brookhaven National Laboratory, a large government laboratory with approximately 3,000 employees, located at the center of Long Island, sixty miles east of New York City. It is famous for a string of discoveries in particle physics, several of which were awarded the Nobel Prize. Most of this research was performed on a large particle accelerator, the Alternate Gradient Synchrotron (AGS). In 1962 Mel Schwartz and his collaborators Leon Lederman and Jack Steinberger discovered a new particle, the muon neutrino. The neutrino that interacts with "muons" was shown to be different from the neutrino that interacts with the electron; thus the muon neutrino is a different particle. This discovery contributed to the modern picture of particle physics, where particles form generations, the muon neutrino belonging to the second generation. Altogether, there are three known generations of particles. Schwartz and his collaborators were awarded the Nobel Prize in 1988 for their discovery. This work was followed shortly after, in 1963, by the discovery of a violation of the "CP" symmetry principle. According to which, the laws of physics would stay the same if all particles were to be replaced by their antiparticles, while all their motions were replaced by their mirror images. It was found that one particle, a neutral K meson, occasionally decays to two pi mesons in violation of that principle. In 1980 the Nobel Prize was awarded to Val Fitch and James Cronin for this discovery. In 1974 a team led by S. C. Ting of MIT discovered the J-particle, which put the quark model of matter on a firm foundation. Their experiment at Brookhaven was the first indication of a new quark, the "charmed" quark. The Nobel Prize was awarded for this discovery two years later, in 1976. A fourth Nobel Prize was awarded for a theoretical discovery. In the summer of 1956, T. D. Lee and C. N. Yang suggested, prior to the experiments, that CP might be violated.

Complementing the big machines, Brookhaven National Laboratory has a physics department similar to the ones at the major universities, in contrast to other large national laboratories that are devoted solely to running large experiments. Thus Brookhaven has an excellent intellectual environment. Most of the activities of the physics department are associated with the large machines, but also a good deal of individual experimental and theoretical research takes place.

I joined a small group of condensed-matter theorists as a postdoctoral fellow in 1974–1976, coming from Denmark, where I had graduated from the Technical University. This fellowship allowed me to work on some of the world's hottest research subjects at that time, critical phenomena associated with equilibrium phase transitions, and organic conducting materials, which can conduct electricity even though they contain no metals such as copper; they are plastic conductors. The work on phase transitions was important for the later work on self-organized criticality because it demonstrated how, under very restrictive conditions, equilibrium systems can exhibit scale-free behavior. The main experiments on organic conductors were performed at Brookhaven's nuclear reactor by Gen Shirane, the world's most accomplished neutron-scatterer, and his collaborators, Alan Heeger and Tony Garito from the University of Pennsylvania. By scattering neutrons off those materials, they obtained information on structural transformations at low temperatures. We were fortunate to have access to the hot experimental data. Vic Emery, who headed the theory group, and I constructed a theory of the most famous of those materials, known as TTF-TCNQ. Contrary to earlier speculations by Heeger and Garito, who had discovered those materials, the transformation was not associated with superconductivity, the exciting capability of certain metals to carry electrical currents without resistance at low temperatures. Our results were reported in the most cited publication in solid state physics of that year. Those were wonderful years.

After the first Brookhaven years, I returned to the University of Copenhagen. Among many other subjects, I became interested in the physics of simple systems with chaotic behavior. Mogens Høgh Jensen, Thomas Bohr (a grandson of Niels Bohr), and I found universal behavior associated with frequency locking of two periodic systems, such as a swing with one natural frequency that is pushed periodically with another frequency. In some sense, self-organized criticality involved a combination of the physics of equilibrium-critical phenomena involving very many particles, which I had studied at Brookhaven, and chaos theory for simple dynamical systems, which I had studied in Copenhagen.

In 1983 I gladly accepted a permanent position in the group. Our group at Brookhaven is a shoestring operation compared with the large machine groups, with only two senior scientists, a couple of postdoctoral research associates, and a number of short- and long-term visitors. Perhaps because of our small size and relative obscurity, we have been able to do basic science, avoiding the relentless pressure to switch to so-called applied science, which in the eyes of science bureaucrats has a perceived immediate payoff. Our agenda is simply to figure out how things work. In the past, we have had the freedom to do whatever we wanted to do, although our budget has been cut every year. Sadly, we have not been able to hire new young scientists for permanent positions for more than a decade. Ironically, this has happened during the most successful period of the group, again because of our invisibility compared with the big machines. Our support is totally unrelated to our scientific accomplishments. In principle, we could sit back, do nothing, and wait for our retirement without any financial consequences.

Contrary to the general public perception, good science today very often comes from small groups consisting of one or two professors and a couple of young collaborators. The dominance of mastodonic science, symbolized by enormous particle accelerators and huge space projects, is over, although there are wonderful exceptions, such as the Hubble telescope. Ideas never occur collectively in the heads of 1,000 individuals. Take a look at some of the most recent Nobel Prize winners in physics: Klaus von Klitzing of Germany for the quantum Hall effect in semiconductors, which involves nothing more complicated than measuring voltages and currents in semiconductors in an electric field; Müller and Bednorz of IBM Zurich for the discovery of high temperature superconductivity; Rohrer and Binnig, also of IBM Zurich, for inventing the tunneling electron microscope; and DeGennes of Paris for theories of polymer physics. All of this is physics at the hundred-thousand-dollar level, carried out by small groups of imaginative scientists left alone to do whatever they wish to do. Indeed, prizes were also awarded for big science throughout those years, but that was mostly to reward big projects based on ideas that were twenty or more years old! Good science is not necessarily expensive science.

Chao Tang came to Brookhaven in 1985 from the University of Chicago, where he had already distinguished himself as a graduate student by some imaginative work on pattern-formation in crystal growth, and on chaos. Kurt Wiesenfeld came from Berkeley where he had been doing similarly impressive work on simple dynamical systems, some of which were showing chaotic behavior. They were holding postdoctoral positions, similar to the one I had in 1974–1976.

Where Does $1/f$ "Noise" Come From?

We became obsessed with the origin of the mysterious phenomenon of $1/f$ noise, or more appropriately, the $1/f$ "signal" that is emitted by numerous sources on earth and elsewhere in the universe. We had endless discussions in the physics coffee room, the intellectual center of Brookhaven. There was a very playful atmosphere, which is crucial for innovative scientific thinking. There would also be a constant stream of visitors passing through and contributing to our research by participating in the discussions, and sometimes by collaborating more directly with us. Good science is fun science.

Most attempts to explain $1/f$ noise were *ad hoc* theories for a single system, with no general applicability, which appeared unsatisfactory to us. Since the phenomenon appears everywhere, we believed that there must be a general, robust explanation. Systems with few degrees of freedom, like the angle and velocity of a single pendulum and equilibrium systems cannot generally show $1/f$ noise or any other complex behavior, since fine-tuning is always necessary. Thus, we came to the conclusion that $1/f$ noise would have to be a cooperative phenomenon where the different elements of large systems act together in some concerted way. Indeed, all the sources of $1/f$ noise were such large systems with many parts. For instance, the fluctuations of the water level of the Nile must be related to the landscape and weather pattern of Africa, which can certainly not be reduced to a simple dynamical system.

One thought was that $1/f$ noise could be related to the spatial structure of matter. Systems in space have many degrees of freedom; one or more degrees of freedom is associated with each point in space. The systems had to be "open," that is, energy had to be supplied from outside, since closed systems in which energy would not be supplied would approach an ordered or disordered equilibrium state without complex behavior. However, at that time there were no known general principles for open systems with many degrees of freedom.

Susan Coppersmith's Dog Model

This was the situation when Susan Coppersmith, a scientist from Bell Laboratories in New Jersey, visited us in late 1986. She had called me a few days before. "I have some new ideas that I am dying to discuss with someone. Can I come and give a presentation to you at Brookhaven? There is nobody here to talk to." How flattering! A little meeting was set up with only three people, Kurt, Chao, and me, in the audience. Sue had been a postdoctoral fellow with us at Brookhaven Lab a few years earlier.

In collaboration with Peter Littlewood, also at Bell Laboratories, she was now working on charge density waves (CDWs) in solid systems. Charge density waves can be thought of as a periodic arrangement of electronic charges, interacting with the regular lattice of atoms in a crystal. She had discovered a simple but remarkable effect.

We can think about CDWs in terms of a simple metaphor. The situation is (very) roughly equivalent to a reluctant dog being pulled along a hilly surface with an elastic leash (Figure 9). At some point the dog will slip, and jump from one bump to the next bump. Because there will still be tension in the string after the jump, the dog will end up at a position near the top of a bump, rather than sliding to the equilibrium position at the bottom of a valley. The dog sits near the top for a while until the tension has built up again to overcome the dog's friction, and the dog will jump again. This can be seen as a trivial example of punctuated equilibria, although with no large events.

This is a simple nonequilibrium open system where energy is supplied from the outside by means of the leash. Actually, a charge density wave can be thought of as a string of particles (dogs), connected with springs, which is pulled across a washboard by means of an external electric field acting as a constant force. Sue's work was based on computer simulations, but together we all came up with a mathematical theory. We studied the situation where the chain would be pulled for some time, and then allowed to relax, and then pulled again. The upshot of the analysis was that after many pulses, most of the particles, just like the dog, would stay near the top of the potential between the pulses. Obviously, particles sitting near the top are much more unstable than particles at the bottom. It would take only a very small push to upset the balance. We called the resulting state "minimally stable." The result of the theory could not possibly be more different from the behavior of equilibrium systems, where all the particles would end up near the bottoms of the valleys in the washboard potential.

Figure 9.   Dog pulled with an elastic string. Every now and then the dog slips from a position near one top, to a position near another top. (Drawing by Ricard Sole).

The basic reason for studying the system was a recent discovery of the phase memory effect by Robert Fleming at Bell Laboratories and George Grüner of UCLA. The positioning of the particles near the tops, in the minimally stable state, beautifully explained that effect.

Indeed, it appeared that it was possible to say something general about open nonequilibrium systems that would distinguish them completely from equilibrium systems. Of course, the resulting configuration has no components of complexity whatsoever, no hints of fractals or $1/f$ noise. But it was the first systematic analysis of large dynamical systems out of equilibrium, once and for all demonstrating the futility of thinking about them in equilibrium terms. New thinking was necessary.

On Coupled Pendulums

Kurt, Chao, and I continued the study of "coupled" systems where many parts interact with one another. Specifically, we looked at a network of coupled torsion pendulums. Figure 10 shows a one-dimensional version where the pendulums are connected along a line. Torsion pendulums can make full rotations around their point of support, not just oscillate around their equilibrium like a clock pendulum. In contrast to previous studies of chaotic behavior in single pendulums, we studied the limit where there were many

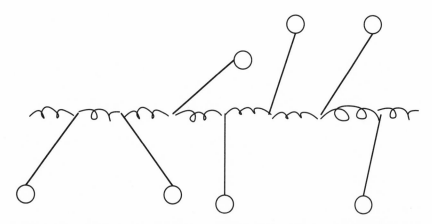

Figure 10.   Coupled pendulums arranged on a chain. At regular intervals, one pendulum, chosen randomly, is pushed so that it makes one revolution. This puts pressure on the neighboring pendulums. We studied a
system where the pendulums were arranged on a two-dimensional grid,
where each pendulum is connected with four neighbors, not two as
shown here.

coupled pendulums. On the computer we put many pendulums on a regular
two-dimensional grid. Neighbor pendulums were connected with springs
like those you find in a clock. Energy was pumped into the system by selecting
one pendulum randomly and pushing it so that it would make one revolution. Because of the network of connected pendulums, this push would put
pressure on the neighbor pendulums by winding up the spring, perhaps forcing one or more of those pendulums also to rotate. The springs were chosen to
be sloppy; it would take several rotations of one pendulum before the force on
the neighboring pendulums would be strong enough to cause a rotation. Our
system was "dissipative." If pushed once and left alone, a pendulum would
make only a single revolution and then stop because of friction. One might
think of the pendulums as rotating in syrup. This contrasts with systems such
as the solar system, which keeps moving forever because it is almost totally
frictionless.

To simplify the calculations we used a representation where we would
keep track only of the number of revolutions, called the winding numbers,
the pendulums would perform; we wouldn't bother with the exact patterns of

rotation. The tension of the springs would depend on the difference in the full number of rotations between neighbor springs. Because of the connecting springs, the winding numbers of neighbor torsion springs cannot differ too much. The dynamics involved only integer numbers, not continuous real numbers; this simplification greatly speeded up the calculations.

The Philosophy of Using Simple Models: On Spherical Cows

Why would we simulate a simple system of oversimplified pendulums instead of a realistic model of something going on in nature? Why don't we do calculations on the real thing?

The answer is simple: there is no such thing as doing calculations on the real thing. One cannot put a frog into the computer and simulate it in order to study biology. Whether we are calculating the orbit of Mercury circling the sun, the quantum mechanics of some molecule, the weather, or whatever, the computer is only making calculations on some mathematical abstraction originating in the head of the scientist. We make pictures of the world. Some pictures are more realistic than others. Sometimes we feel that our modeling of the world is so good that we are seduced into believing that our computer contains a copy of the real world, so that real experiments or observations are unnecessary. I have fallen into that trap when sitting too long in front of the computer screen. Obviously, if we want our calculation to produce accurate quantitative results, such as on the weather, or accurate predictions, such as of the rate of global warming, the demands are much more stringent than when only qualitative behavior is asked for. This is true not only for computer modeling but also for pen-and-paper analytical calculations like those performed by the geneticists in the 1930s. The absence of computers put even more severe limitations on the type of calculations that could be done. When scientists in the past made theories of evolution, for example, they made theories of simple models of evolution. Instead of calculating the probabilities of reproduction and survival in the real world, all of this information might be condensed into a single abstract number called fitness, which would enter the calculation. We are always dealing with a model of the system, although some scientists would like us to believe that they are doing calculations on the real system when they ask us to believe their results, whether it be on global warming or the world economy.

The large dynamical systems that we are interested in, like the crust of the earth, are so complicated that we cannot hope to make accurate enough calculations to predict what will happen next, even if we join the forces of all the computers in the world. We would have to construct a full-sized model of California in order to predict where and when the next large earthquake would take place. This is clearly a losing strategy!

The physicist's approach is complementary to that of an engineer, who would try to add as many features to the model as are necessary to provide a reliable calculation for some specific phenomenon. The physicist's agenda is to understand the fundamental principles of the phenomenon under investigation. He tries to avoid the specific details, such as the next earthquake in California. Before asking how much we have to add to our description in order to make it reproduce known facts accurately, we ask how much we can throw out without losing the essential qualitative features. The engineer does not have that luxury! Our strategy is to strip the problem of all the flesh until we are left with the naked backbone and no further reduction is possible. We try to discard variables that we deem irrelevant. In this process, we are guided by intuition. In the final analysis, the quality of the model relies on its ability to reproduce the behavior of what it is modeling!

Thus, how would we physicists make a suitable model of, say, biological evolution? The biologist might argue that since there is sexual reproduction in nature, a theory of evolution must necessarily include sex. The physicist would argue that there was biology before there was sex, so we don't have to deal with that. The biologist might point out that there are organisms with many cells, so we must explain how multicellular organisms developed. The physicist argues that there are also single-cell organisms, so we can throw multicellular organisms out! The biologist argues that most life is based on DNA, so that should be understood. The physicist emphasizes that there is simpler life based on RNA, so we don't have to deal with DNA. He might even argue that there must have been a simpler reproductive chemistry before RNA, so that we don't have to deal with that either, and so on. The trick is to stop the process before we throw out the baby with the bathwater. Once we have identified the basic mechanisms from the simple models, we leave it to others to put more meat on the skeleton, to add more and more specific details, if one so wishes, to check whether or not more details modify the results.

In our particular study, the underlying philosophy is that general features, such as the appearance of large catastrophes and fractal structure, cannot be sensitive to the particular details. This is the principle of universality. We hope that important features of large-scale phenomena are shared between seemingly disparate kinds of systems, such as a network of interacting economics agents, or the interactions between various parts of the crust of the earth. This hope is nourished by the observation of the ubiquitous empirical patterns in nature—fractals, $1/f$ noise, and scaling of large events among them—discussed in Chapter 1. Since these phenomena appear everywhere, they cannot depend on any specific detail whatsoever.

Universality is the theorist's dream come true. If the physics of a large class of problems is the same, this gives him the option of selecting the *simplest* possible system belonging to that class for detailed study. One hopes that a system is so simple that it can be studied effectively on a computer, or maybe laws of nature can be derived by mathematical analysis, with pen and paper, from that stripped-down description or model. Simple models also serve to strengthen our intuition of what goes on in the real world by providing simple metaphoric pictures.

The concept of universality has served us well in the past. It has scored a couple of spectacular successes in recent years. Wilson's theory of phase transitions for which he was awarded the Nobel Prize proved its universality by demonstrating that the basic properties of a system near a phase transition had nothing to do with the microscopic details of the problem. It doesn't matter whether we are dealing with a liquid–gas transition, a structural transition where a crystal deforms, or a magnetic transition where the little magnets or spins start pointing in the same direction. Wilson's calculations were based on the Ising model, the simplest possible model of a phase transition, and they agreed with experiments on much more complicated real systems, such as magnets and fluids.

Similarly, Feigenbaum's studies of the transition to chaos was based on a "map" that can only be seen as a caricature of a real predator-prey ecological system. I don't think that either Feigenbaum or May ever claimed that the map describes anything in real biology. Feigenbaum argued that near the transition to chaos the dynamics had to be the same for all systems undergoing a transition to chaos through an infinite sequence of bifurcations at which the periodicity would be doubled. The contrast between the simplicity of the model, and the depth of the resulting behavior is astonishing. Although Feigenbaum's theory was based on a grossly oversimplified model, experiments on many kinds of complicated systems have beautifully confirmed it. In particular, Albert Libchaber in Paris showed that a liquid with rotating convective rolls would undergo a series of transitions and ultimately goes to a chaotic state following Feigenbaum's law. Another simpler example is the swing, or pendulum, being pushed repeatedly at a constant rate, which I studied with Bohr and Jensen. Again, real-world behavior, representing real measurable quantities, could be predicted from simple model calculations. The phenomenon is quite universal.

Thus, the scientific process is as follows: We describe a class of phenomenon in nature by a simple mathematical model, such as the Feigenbaum map. We analyze the model either by mathematical analytical means, with pen and paper, or by numerical simulations. There is no fundamental difference between these two approaches; they both serve to elucidate the consequences (predictions) of the simple model. Often, however, simulations are easier than mathematical analysis and serve to give us a quick look at the consequences of our model before starting analytical considerations. Computational physics does not represent a "third" way of doing science, in addition to experiments and theory. There is no fundamental difference, except that it is more convenient, compact, and elegant to have a closed mathematical formula rather than a computer program. We then compare the findings with experiments and observations. If there is general agreement, we have discovered new laws of nature operating at a higher level. If not, we haven't. The beauty of the model can be measured as the range between its own simplicity and the complexity of the phenomena that it describes, that is, by the degree to which it has allowed us to condense our description of the real world.

Without the concept of universality we would be in bad shape. There would be no fundamental "emergent" laws of nature to discover, only a big mess. Of course, we have to demonstrate that our models are robust, or insensitive to changes, in order to justify our original intuition. If, unfortunately, it turns out that they are not, we are back to the messy situation where detailed engineering-type models of the highly complex phenomena is the only possible approach—the weatherman's approach.

The obsession among physicists to construct simplified models is well illustrated by the story about the theoretical physicist asked to help a farmer raise cows that would produce more milk. For a long time, nobody heard from him, but eventually he emerged from hiding, in a very excited state. "I now have figured it all out," he says, and proceeds to the blackboard with a piece of chalk and draws a circle. "Consider a spherical cow. . . ." Here, unfortunately, it appears that universality does not apply. We have to deal with the real cow.

The Pendulums Become Critical

This is why we were finding ourselves doing computer simulations on something as esoteric as networks of coupled pendulums—and not realistic models of earthquakes or whatever. If the reader has difficulties visualizing the system of coupled pendulums, so much the better—it will only serve to illustrate the value of having good metaphors. The pendulums are not good enough metaphors. We too had great difficulty grasping what was going on, and it was still too messy.

If the pendulums were pushed in random directions, one at a time, nothing interesting would happen. Most of the pendulums would be near the down position. However, we realized that if we always pushed the pendulums in the same direction, say clockwise, there would be an increased tendency for the pendulums to affect each other. The springs connecting the pendulums would slowly be wound up and store energy. As the process of pushing a single pendulum at a time continued, more and more pendulums would stay near the upward position rather than the downward position. Because of the increased instability of the pendulums, there would be chain reactions caused by a domino effect. Pushing a single pendulum might cause others to rotate. How far would this domino process continue? Obviously, if we started from the position where all the springs were relaxed, there would be no way that pushing a single pendulum once would cause other pendulums to rotate. But suppose the process of pumping up the pendulums went on for a very long time. What would set the limit of the chain reaction? What would be the natural scale of the disturbances? How many pendulums could be turned by a single push?

The idea popped up that maybe there was no limit whatsoever! It appeared that there was essentially nothing in the system that could possibly define a limit! Maybe, even if the system was dissipative with lots of friction, the constant energy supply from pushing the pendulums might eventually drive the system to a state where once a single pendulum started rotating somewhere, there would be enough stored energy to allow a chain reaction to go on forever, limited only by the large total number of pendulums?

Chao Tang programmed this into a computer. He chose a small system with pendulums on a grid of size 50 by 50, a total of 2,500 pendulums. Each pendulum was connected with its four neighbors, in the up, down, left, and right directions. Starting from having all pendulums in the down direction, one arbitrary pendulum would be wound up by one revolution. This would put more pressure on the neighbors. Then another pendulum would be chosen, and so on. For a while there were only single rotations, but at some point one spring would be wound up enough to trigger another pendulum to rotate. Continuing further, at some point there would be enough energy stored in the springs that there would be large chain reactions, where one pendulum would trigger the next by a domino effect. This process is called an avalanche. The avalanches would become bigger and bigger. Eventually, after thousands of events, they would grow no further. As the simulation continued, there would be a stream of avalanches, some small, some intermediate, and a few big.

We measured how many avalanches there were of each size, just like the earthquake scientists had measured how many earthquakes there were of each magnitude. The size of an avalanche was measured as the total number of rotations following a single kick. There were many more small ones than large ones. Figure 13 shows the resulting histogram. The x-axis shows the size of the avalanches. The y-axis shows how many avalanches there were of that size. We used log-log plots, just like Johnston and Nava in Figure 2, and Zipf in Figure 8.

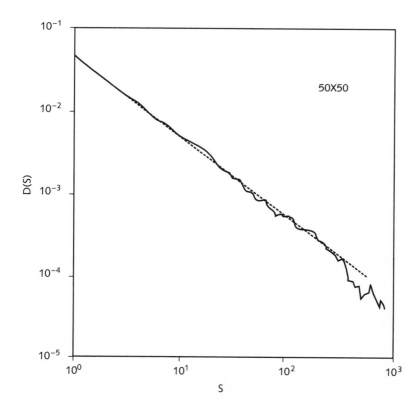

Figure 11. Size distribution of avalanches in systems of coupled pendulums or, equivalently, in the sandpile model. The figure shows how many avalanches there are of each size, on a logarithmic plot. The distribution is a power law with exponent 1.1. This is our very first plot. By performing longer simulations on bigger system one can extend the range of the power law.

Our data fall approximately on a straight line, which indicates that the number of avalanches of size s is given by the simple power law

$$N(s) = s^{-\tau}$$

where the exponent τ, defined as the slope of the curve, is approximately equal to 1.1. The pendulums obeyed the Gutenberg–Richter power law for earthquakes! At the lower end, the straight line is limited by the fact that no avalanche can be smaller than one pendulum rotation. At the upper end, there is a cutoff because no avalanche can be bigger than one with all the pendulums rotating. The scattering of points around the straight line are statistical fluctuations, just like in real experiments. Some points are above the line, some below. If we let the simulation continue longer and longer, these fluctuations become smaller and smaller, just like the ratio of sixes you get when you throw a dice will converge toward 1/6 as the number of throws increases.

The system had become "critical"! There were avalanches of all sizes just as there were clusters of all sizes at the "critical" point for equilibrium phase transitions. But no tuning was involved. We had just blindly pushed the pendulums. There is no temperature to regulate, no λ parameter to change. The simple behavior of the individual elements following their own simple local rules had conspired to create a unique, delicately balanced, poised, global situation in which the motion of any given element might affect any other element in the system. The local rule was simply a specification of the total number, n, of revolutions the four neighbors should perform, to induce a single revolution of a given pendulum. *The system had self-organized into the critical point without any external organizing force.* Self-organized criticality (SOC) had been discovered. It was as if some "invisible hand" had regulated the collection of pendulums precisely to the point where avalanches of all sizes could occur. The pendulums could communicate throughout the system.

Once the poised state has been reached, the "criticality" is similar to that of a nuclear chain reaction. Suppose you have a collection of radioactive atoms emitting neutrons. Some of those neutrons might become absorbed by other atoms, causing them to emit neutrons of their own. A single neutron leads to an avalanche. If the concentration of fissionable atoms is low, the chain reaction will die out very soon. If the concentration is high, there will be a nuclear explosion similar to that in an atomic bomb. At a unique critical concentration there will be avalanches of all sizes, all of which will eventually stop. Again, one has to "tune" nuclear chain reaction by choosing precisely the correct amount of radioactive material to make it critical. In nuclear reactors this tuning is very important and is carried out by inserting neutron-absorbing graphite rods. In general the reactor is not critical. There is absolutely no self-organization involved in a nuclear chain reaction, so in this all-important aspect the situation is entirely different.

Fermi's team, achieving criticality at their nuclear reactor in Chicago in 1940, could not have been more excited than we were. Criticality, and therefore complexity, can and will emerge "for free" without any watchmaker tuning the world.

the sandpile paradigm

The importance of our discovery of the coupled-pendulums case of self-organized criticality was immediately obvious to us. An open dissipative system had naturally organized itself into a *critical* scale-free state with avalanches of all sizes and all durations. The statistics of the avalanches follow the Gutenberg–Richter power law. There were small events and large events following the same laws. We had discovered a simple model for complexity in nature.

The variability that we observe around us might reflect parts of a universe operating at the self-organized critical state. While there had been indications for some time that complexity was associated with criticality, no robust mechanism for achieving the critical state had been proposed, nor had one been demonstrated by actual calculation on a real mathematical model. Of course, this was only the beginning. For instance, we still had to show that the activity has an $1/f$-like signal, and that the resulting organization had a fractal geometrical structure. We were only at the beginning.

Perhaps our ultimate understanding of scientific topics is measured in terms of our ability to generate metaphoric pictures of what is going on. Maybe understanding *is* coming up with metaphoric pictures. The physics of our messy system of pendulums was far from transparent. Our intuition was poor. A couple of months after the discovery, it struck us that there was a simpler picture that could be applied to our self-organized critical dynamics. By a change of language the rotating pendulums could be describing toppling grains of sand in a pile of sand (Figure 1). Instead of counting revolutions of pendulums, we would count toppling grains at some position in the pile. Although the mathematical formulation was exactly the same for the sand model as for the pendulum model, the sand picture led to a vastly improved intuitive understanding of the phenomenon. Sandpiles are part of our everyday experience, as any child who has been playing on the beach knows. Rotating coupled pendulums are not. In a mysterious way, the physical intuition based on the sandpile metaphor leads to better understanding of the behavior of a purely mathematical model. Usually we achieve physical understanding from mathematical analysis, not the other way around.

But before discussing the mathematical formulation of our model, let us recall the sandpile experiment in Chapter 1. Consider a flat table, onto which sand is added slowly, one grain at a time. The grains might be added at random positions, or they may be added only at one point, for instance at the center of the table. The flat state represents the general equilibrium state; this state has the lowest energy, since obviously we would have to add energy to rearrange the sand to form heaps of any shape. If we had used water, the system would always return to the flat ground state as the water would simply run off the edge of the table. Because the grains tend to get stuck due to static friction, the landscape formed by the sand will not automatically revert to the ground state when we stop adding sand.

Initially, the grains of sand will stay more or less where they land. As we continue to add more sand, the pile becomes steeper, and small sand slides or avalanches occur. The grain may land on top of other grains and topple to a lower level. This may in turn cause other grains to topple. The addition of a single grain of sand can cause a local disturbance, but nothing dramatic happens to the pile. In particular, events in one part of the pile do not affect sand grains in more distant parts of the pile. There is no global communication within the pile at this stage, just many individual grains of sand.

As the slope increases, a single grain is more likely to cause other grains to topple. Eventually the slope reaches a certain value and cannot increase any further, because the amount of sand added is balanced on average by the amount of sand leaving the pile by falling off the edges. This is called a stationary state, since the average amount of sand and the average slope are constant in time. It is clear that to have this average balance between the sand added to the pile, say, in the center, and the sand leaving along the edges, there must be communication throughout the entire system. There will occasionally be avalanches that span the whole pile. This is the self-organized critical (SOC) state.

The addition of grains of sand has transformed the system from a state in which the individual grains follow their own local dynamics to a critical state where the emergent dynamics are global. In the stationary SOC state, there is one complex system, the sandpile, with its own emergent dynamics. The emergence of the sandpile could not have been anticipated from the properties of the individual grains.

The sandpile is an open dynamical system, since sand is added from outside. It has many degrees of freedom, or grains of sand. A grain of sand landing on the pile represents potential energy, measured as the height of the grain above the table. When the grain topples, this energy is transformed into kinetic energy. When the toppling grain comes to rest, the kinetic energy is dissipated, that is, transformed into heat in the pile. There is an energy flow through the system. The critical state can be maintained only because of energy in the form of new sand being supplied from the outside.

The critical state must be robust with respect to modifications. This is of crucial importance for the concept of self-organized criticality to have any chance of describing the real world; in fact, this is the whole idea. Suppose that after the same system has reached its critical stationary state we suddenly start dropping wet sand instead of dry sand. Wet sand has greater friction than dry sand. Therefore, for a while the avalanches would be smaller and local. Less material will leave the system since the small avalanches cannot reach the edge of the table. The pile becomes steeper. This, in turn, will cause the avalanches to grow, on average. Eventually we will be back to the critical state with system-wide avalanches. The slope at this state will be higher than the original ones. Similarly, if we dry the sand, the pile will sink to a more shallow shape by temporarily shedding larger avalanches. If we try to prevent avalanches by putting local barriers, "snow" screens, here and there, this would have a similar effect: for a while the avalanches will be smaller, but eventually the slope will become steep enough to overcome the barriers, by forcing more sand to flow somewhere else. The physical appearance of the pile changes, but the dynamics remain critical. The pile bounces back to a critical state when we try to force it away from the critical state.

The Sandpile Model

We have defined the physics, but so far everything is simply a product of imagination, mixed with some intuition from actual experience. How do we go from here to make a representation, a model, that reproduces these features? The sandpile model that Kurt, Chao, and I studied is easy to define and simulate on the computer. It is so simple that readers who possess some computer literacy can set one up on their own PCs. Readers who do not play with computers can make a mechanical representation using Lego blocks.

The table where the sand is dropped is represented by a two-dimensional grid. At each square of the grid, with coordinates (x,y), we assign a number $Z(x,y)$, which represents the number of grains present at that square. For a table of size $L = 100$, the coordinates x and y are between 1 and 100. The total number of sites is $L \times L$. We are using "theoretical physicist's sand," with ideal grains that are regular cubes of size 1, which can be stacked neatly on top of one another, not the irregular complicated ones that you find on the beach.

The addition of a grain of sand to a square of the grid is carried out by choosing one site randomly and increasing the height Z at that site by 1:

$$Z(x,y) \rightarrow Z(x,y) + 1.$$

This process is repeated again and again. To have some interesting dynamics, we apply a rule that allows a grain of sand to shift from one square to another, a "toppling rule." Whenever the height Z exceeds a critical value Z_{cr} that may arbitrarily be set, say, to 3, one grain of sand is sent to each of the four neighbors. Thus, when Z reaches 4, the height at that site decreases by four units,

$$Z(x,y) \to Z(x,y) - 4$$

for $Z(x,y) > Z_{cr}$, and the heights Z at the four neighbor sites go up by one unit,

$$Z(x \pm 1, y) \to Z(x \pm 1, y) + 1, Z(x, y \pm 1) \to Z(x, y \pm 1) + 1.$$

The toppling process is illustrated in Figure 12. If the unstable site happens to be at the boundary, where x or y is 1 or 100, the grains of sand simply leave the system; they fall off the edge of the table and we are not concerned with them any longer.

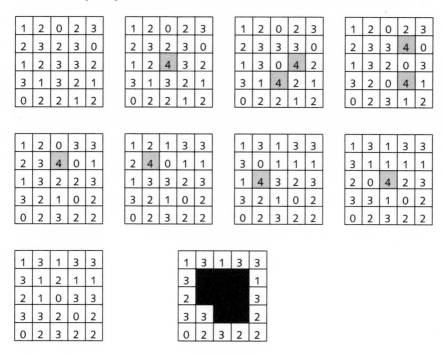

Figure 12. Illustration of toppling avalanche in a small sandpile. A grain falling at the site with height 3 at the center of the grid leads to an avalanche composed of nine toppling events, with a duration of seven update steps. The avalanche has a size $s = 9$. The black squares indicate the eight sites that toppled. One site toppled twice.

These few simple equations completely define our model. No mathematics more complicated than adding and subtracting numbers between 1 and 4 is needed. Nevertheless, the consequences of these rules are horrifyingly complicated, and can certainly not be deduced from a simple inspection of the equations, which represent the local dynamics of each of our sand grains. We follow the general procedure outlined in Chapter 2, and start studying the model by direct computer simulations.

This physicists' sandpile is a gross oversimplification of what really happens. First, real grains have different sizes and shapes. The instabilities in a real sandpile occur not only at the surface but also through the formation of cracks in the bulk. The toppling depends on how the individual grains lock together. Once a grain is falling, its motion is determined by the gravity field, which accelerates the grain, and the interaction with other grains, which tends to decelerate the motion. Stopping the motion depends on many factors, such as the shape of the grains it bumps into and its velocity at that point, and not just the height or slope of the pile at the neighbor points. One could go on and on with objections like this. One quickly realizes that it is a losing strategy to make a realistic model of the sandpile, which at first glance might have seemed a reasonably simple object. So why is the model acceptable at all? Its validity is based on the intuition that the model contains the essential physics, namely that grains interact and may cause each other to topple. That this is indeed correct can be justified (or falsified) only a posteriori by comparing with experiments.

Second, we are not particularly interested in sand. We hope that the sand dynamics that we observe are general enough that they can be applied to a much larger class of phenomena.

Peter Grassberger, a computational physicist at the University of Wuppertal, Germany has come up with an amusing representation of the model. He asks us to think about a large office where bureaucrats sit at tables organized in rows (Figure 13). Every now and then a piece of paper from the outside enters the desk of a random bureaucrat. He does not deal with it until he finds too many pieces of paper on his desk. He then sends one piece of paper to each of his four neighbors. Everybody follows this rule, except those who are placed along the walls, who simply throw the paper out the window. Jumping

Figure 13. Office version of the sandpile model. At regular intervals a piece of paper lands on the desk of a random bureaucrat. When a bureaucrat finds four or more sheets of paper on his desk he sends one sheet to each of his neighbors, or out the window. (Courtesy of Peter Grassberger.)

forward a little bit, we shall see that a single piece of paper entering the office can lead to a bureaucratic catastrophe where millions of transfers of paper take place (if the office is large enough!). Each bureaucrat may perform many transactions within such an avalanche.

In the beginning of the process, where all the heights are low, there are no unstable sites. All sites have Z less than 3, so the sand stays precisely where it happens to land. After many steps of adding a single grain to a square of the grid, the height somewhere must necessarily exceed 3, and we have the first toppling event. It is unlikely that the height at any of the four neighbor squares exceeds 3 this soon, so there will be no further activity of toppling grains. As the process continues, it becomes more likely that at least one of the neighbors will reach its critical height, so the first toppling event induces a second event. One toppling event leads to the next, like falling dominos. As more sand is added, there will be bigger and bigger landslides, or avalanches, although there will still also be small ones.

Figure 12 shows a sequence of toppling events in a very small system. The numbers in the squares represent the heights. A grain of sand lands on a site with height 3, causing that site to topple. Two of the neighbor sites had $Z = 3$, so those two sites topple next, at the second time step, sending a total of eight grains to their neighbors, including two grains to the original site. Eventually the system comes to rest. We notice that there were precisely nine topplings, so that avalanche had size $s = 9$. We also monitor the total duration, that is, the number of update steps, $t = 7$, of that avalanche.

Eventually the entire sandpile enters into a stationary state where the average height of all sites does not increase further. The average height is somewhere between 2 and 3. The pile can never reach the highest possible stable state, where all the heights are 3, since long before that simple state is reached the pile has broken down due to large avalanches. We can monitor this by counting the total number of grains in the pile at all times. In the stationary state, most avalanches are small and do not reach the edge, so they cause the pile to grow. This is precisely compensated by fewer, and generally larger, avalanches that reach the edge and cause many grains of sand to leave the pile.

Plate 1a shows a configuration in the stationary state, just after the completion of an avalanche for a very large pile. Here, instead of the numbers, a color code is used. Red is $Z = 3$, blue is $Z = 2$, green is $Z = 1$, and gray is $Z = 0$. The picture looks like a big mess, with no organized structure whatsoever. But nothing can be further from the truth. The pile has organized itself into a highly orchestrated, susceptible state through the process of repeatedly adding sand and having avalanches travel through the pile again and again.

We can realize the intricate properties of the configuration of sand, not by directly inspecting the colors but by dropping one more grain of sand. If a "red" site is hit, this triggers an avalanche. Plate 1b shows what has happened after a few time steps. The light blue area represents all the grains that have fallen. The yellow and white spots represent active sites that are about to topple, where $Z > 3$. The next picture shows the situation a little later, where the avalanche has covered a larger area. Eventually the avalanche comes to a stop after approximately half the sites in the pile have toppled at least once. Most sites have actually toppled several times. The particular configuration at the end of the avalanche is very different than the one we started out with.

This was a very big avalanche. More often than not the avalanches are smaller. We now follow the same procedures as the geophysicists when making statistics of earthquakes. By successively adding sand after each avalanche has stopped we generate a large series of avalanches, say 1 million avalanches. We then make a "synthetic" earthquake catalog by counting how many avalanches there are of each size. The "magnitude" of avalanches is the logarithm of the size of the avalanche. As usual, we plot the logarithm of the number of avalanches of a given magnitude versus that magnitude.

The number of avalanches of each size for a system of linear size 50 is plotted in Figure 11 on p. 47, which shows data from our very first sandpile. The straight line indicates that the avalanches follow the Gutenberg–Richter power law, just like the real earthquakes in Figure 2, although the slopes are different. We do not have to wait millions of years to generate many earthquakes, so our statistical fluctuations are smaller than those for earthquakes, where we must deal with the much smaller number that nature has generated for us. The exponent τ of the power law, that is, the slope of the curve in Figure 11, was measured to be approximately 1.1. The power law indicates that the stationary state is critical. We conclude that the pile has self-organized into a critical state.

One can show, by analyzing the geometry of the sandpile, that the profile of the sandpile is a fractal, like Norway's coast. The avalanches have carved out fractal structures in the pile.

The power law also indicates that the distribution of avalanches follows Zipf's law. Instead of plotting how many avalanches there are of each size, we could equally well plot how large the biggest avalanche was (the avalanche of "rank" 1), how large the second biggest avalanche, of rank 2 was, how large the tenth biggest avalanche was, and so on, precisely the same way that Zipf plotted the ranking of cities. This is just another way of representing the information from the original power law. The straight line shows that the sandpile dynamics obey Zipf's law.

Our simple model cannot by any stretch of the imagination represent the formation of real cities in a human society or the process by which James Joyce wrote *Ulysses*, where we are dealing with humans, not sand grains. Nevertheless, one might speculate that Zipf's law indicates that the world population has organized itself into a critical state, where cities are formed by avalanches of human migrations.

We had to check that the criticality is robust with respect to modifications of the model. The power law should prevail no matter how we modify the sandpile. We tried a long sequence of different versions. Instead of having the same critical height equal to 3, a version where the critical height varies from site to site was tried. Snow screens were simulated by preventing sand from falling between certain neighbor sites, selected randomly, by having the sand arranged on a triangular grid instead of the square grid. We also tried adding grains of different sizes, that is, we increased Z not by unity when grains are falling but by some random number between 0 and 1. We massaged the model so that a random amount of sand topples when the site becomes unstable. We selected the sites to which the sand would topple in a random way, and not to the nearest neighbors. In all cases, the pile organized itself into a critical state with avalanches of all sizes. The criticality was unavoidable.

One might speculate that the criticality is caused by the randomness of the way that the system is driven—we add new grains at random positions. In fact, this is not important at all. We can drive the system in a deterministic way with no randomness whatsoever, with all information about the system at all times encoded in the initial condition: let the Zs represent a real variable instead of an integer one. Start with a configuration where all the Zs are subcritical, that is, less than 4. Increase all Zs at a very small rate. This corresponds to tilting the sandpile slowly. At some point, one Z will become unstable and topple according to the rule defined above, and a chain reaction is initiated. The process is continued ad infinitum; there will eventually be a balance between the rate of changing the slope and the rate of sand falling off the edges. We get the same power law distribution as before. Since the whole history of the pile in this case was contained in the initial condition, the phenomenon of SOC is essentially a deterministic phenomenon, just like the chaos studied by Feigenbaum.

The fact that the randomness of adding sand does not affect the power law indicates that the randomness is irrelevant for the complex behavior we are observing. This fact is important to realize when studying much more complicated systems. Economics deals with the more or less random behavior of many agents, whose minds were certainly not made up at the beginning of history. Nevertheless, this randomness does not preclude the system's evolving to the delicate critical state, with well-defined statistical properties. This is a fascinating point that is difficult to grasp. How can a system evolve to an organized state despite all the obvious randomness in the real world? How can the particular configuration be contingent on minor details, but the criticality totally robust?

Life in the Sandpile World

The dynamics of the nonequilibrium critical state could hardly be more different than the quiet dynamics of a flat beach. How would a local observer experience the situation? During the transient stage, when the sandpile was relatively shallow, his experience would be monotonous. Every now and then there would be a small disturbance passing by, when a few grains topple in the neighborhood. If we drop a single grain of sand at one place instead of another, this causes only a small local change in the configuration. There is no means by which the disturbance can spread system-wide. The response to small perturbations is small. In a noncritical world nothing dramatic ever happens. It is easy to be a weather (sand) forecaster in the flatland of a noncritical system. Not only can he predict what will happen, but he can also understand it, to the limited extent that there is something to understand. The action at some place does not depend on events happening long before at faraway places. *Contingency* is irrelevant.

Once the pile has reached the stationary critical state, though, the situation is entirely different. A single grain of sand might cause an avalanche involving the entire pile. A small change in the configuration might cause what would otherwise be an insignificant event to become a catastrophe. The sand forecaster can still make short time predictions by carefully identifying the rules and monitoring his local environment. If he sees an avalanche coming, he can predict when it will hit with some degree of accuracy. However, he cannot predict when a large event will occur, since this is contingent on very minor details of the configuration of the entire sandpile. The relevance of contingency in the self-organized critical state was first noted by Maya Paczuski, then a research fellow in our group, who suggested that the massive contingency in the real world could be understood as a consequence of self-organized criticality.

The sand forecaster's situation is similar to that of the weatherman in our complex world: by experience and data collection he can make "weather" forecasts of local grain activity, but this gives him little insight into the "climate," represented by the statistical properties of many sand slides, such as their size and frequency.

Most of the time things are completely calm around him, and it might appear to him that he is actually living in a stable equilibrium world, where nature is in balance. However, every now and then his quiet life is interrupted by a punctuation—a burst of activity where grains of sand keep tumbling around him. There will be bursts of all sizes. He might be tempted to believe that he is dealing with a local phenomenon since he can relate the activity that he observes to the dynamical rules of the sand toppling around him. But he is not; the local punctuation that he observes is an integrated part of a global cooperative phenomenon.

Parts of the critical system cannot be understood in isolation. The dynamics observed locally reflect the fact that it is part of an entire sandpile. If you were sitting on a flat beach instead of a sandpile, the rules that govern the sand are precisely the same, following the same laws of physics, but history has changed things. The sand is the same but the dynamics are different. The ability of the sand to evolve slowly is associated with its capability of recording history. Sand may contain memory; one can write letters in the sand that can be read a long time later. This cannot happen in an equilibrium system such as a dish of water.

In the critical state, the sandpile is the functional unit, not the single grains of sand. No reductionist approach makes sense. The local units exist in their actual form, characterized for instance by the local slope, only because they are a part of a whole. Studying the individual grains under the microscope doesn't give a clue as to what is going on in the whole sandpile. Nothing in the individual grain of sand suggests the emergent properties of the pile.

The sandpile goes from one configuration to another, not gradually, but by means of catastrophic avalanches. Because of the power law statistics, most of the topplings are associated with the large avalanches. The much more frequent small avalanches do not add up to much. Evolution of the sandpile takes place in terms of revolutions, as in Karl Marx's view of history. Things happen by revolutions, not gradually, precisely because dynamical systems are poised at the critical state. Self-organized criticality is nature's way of making enormous transformations over short time scales.

In hindsight one can trace the history of a specific large avalanche that occurred. Sand slides can be described in a narrative language, using the methods of history rather than those of physics. The story that the sand forecaster would tell us goes something like this:

"Yesterday morning at 7 A.M., a grain of sand landed on site A, with coordinates $(5,12)$. This caused a toppling to site B at $(5,13)$. Since the grain of sand resting at B was already near the limit of stability, this caused further topplings to sites C, D, and E. We have carefully monitored all subsequent topplings, which can easily be explained and understood from the known laws of sand dynamics, as expressed in the simple equations. Clearly, we could have prevented this massive catastrophe by removing a grain of sand at the initial triggering site. Everything is understood."

However, this is a flawed line of thinking for two reasons. First, the fact that this particular event led to a catastrophe depended on the very details of the structure of the pile at that particular time. To predict the event, one would have to measure everything everywhere with absolute accuracy, which is impossible. Then one would have to perform an accurate computation based on this information, which is equally impossible. For earthquakes, we would have to know the detailed fault structure and the forces that were acting on those faults everywhere in a very large region, like California. Second, even if we were able to identify and remove the triggering grain, there would sooner or later be another catastrophe, originating somewhere else, perhaps with equally devastating consequences.

But most importantly, *the historical account does not provide much insight into what is going on, despite the fact that each step follows logically from the previous step.* The general patterns that are observed even locally, including the existence of catastrophic events, reflect the fact that the pile had evolved into a critical state during its entire evolutionary history, which took place on a much longer time scale than the period of observation. The forecaster does not understand why the arrangement of grains happened to be precisely such that it could accommodate a large avalanche. Why couldn't all avalanches be small?

There is not much that an individual can do to protect himself from these disasters. Even if he is able to modify his neighborhood by flattening the pile around him, he might nevertheless be swept away by avalanches from far away, through no fault of his own. Fate plays a decisive role for the sandpile inhabitant. In contrast, the observer on the flat noncritical pile can prevent the small disasters by simple local measures, since he needs information only about his neighborhood in order to make predictions, assuming that he has information on the arrival of grains to the pile. It is the criticality that makes life complicated for him.

The sandpile metaphor has reached well beyond the world of physicists thinking about complex phenomena; it contains everything—cooperative behavior of many parts, punctuated equilibrium, contingency, unpredictability, fate. It is a new way of viewing the world Vice President Al Gore says in his book *Earth in the Balance:*

> The sand pile theory—self-organized criticality—is irresistible as a metaphor; one can begin by applying it to the developmental stages of human life. The formation of identity is akin to the formation of the sand pile, with each person being unique and thus affected by events differently. A personality reaches the critical state once the basic contours of its distinctive shape are revealed; then the impact of each new experience reverberates throughout the whole person, both directly, at the time it occurs, and indirectly, by setting the stage for future change. . . . One reason I am drawn to this theory is that it has helped me understand change in my own life.

Maybe Gore is stretching the point too far. On the other hand, perhaps even the most complicated phenomena on earth—humans with brains and personality—do reflect part of a world operating at the critical state. We shall return to these issues in the context of biological evolution and brain function in later chapters.

Can We Calculate the Power Laws with Pen and Paper?

The sandpile model is utterly simple to describe. It takes only a couple of lines of text to define the model completely. Why do we have to go through the computer simulation? The computer calculation does not prove anything in the mathematical sense. Can't we make a simple pen-and-paper calculation that will tell us what will happen without the simulation? For instance, can we calculate the exponent τ for the distribution of avalanches? The model is so simple and transparent that one would expect to be able to calculate everything. For other complicated phenomena, like the transition to chaos, or phase transitions in equilibrium systems, scientists like Feigenbaum and Wilson were eventually able to create beautiful analytical theories providing deep insight into the origins.

Surprisingly, we cannot! Some of the best brains in mathematical physics have been working on the problem, including Mitch Feigenbaum and Leo Kadanoff of the University of Chicago, and Itamar Procaccia of the Weizmann Institute in Israel. Together with a couple of very bright graduate students, Chhabra and Kolan, they considered a model that is even simpler than ours: the grains were arranged in a one-dimensional pile where sand was stacked on a line, not a two-dimensional plane. The model self-organizes to the critical point, but no analytical results could be derived. For instance, they were unable to prove that the avalanches follow a power law despite a monumental effort published in a long article in *Physical Review*.

In a very beautiful mathematical theory, the physicist Deepak Dhar from the Tata Institute at Bombay was able to calculate some properties; he calculated how many possible sandpile configurations exist in the critical state. He also constructed an algorithm that allows us to check whether a specific configuration, like the one shown in Plate 1, represents a configuration that can be found in the stationary state of the pile, or whether, conversely, it is a transient state representing a sandpile that has not yet reached its stationary state. But he was not able to calculate the all-important exponent τ or to prove that the stationary state has power law distribution of avalanches.

The mathematics is prohibitively difficult. But how can it be otherwise? We deal with the most complex phenomena in nature, involving a slow buildup of information through a long history; why should we necessarily expect a simple mathematical formula to describe this state?

The model is simple, but nevertheless too difficult for theoretical physicists and mathematicians to analyze efficiently. At least so far no one has been able to deal with it satisfactorily. This situation might have dampened some enthusiasm.

In a subsequent chapter we shall see that for some other models we can achieve a good deal of analytical insight. We can understand the basic nature of the self-organization process. We can relate some exponents to other exponents. In some simplified but even more artificial models where sand topples to random positions, one can calculate the exponents, and explicitly show that the pile self-organizes to the critical state.

We shall also see that there are other models that describe surface growth, traffic, and biological evolution, where pen-and-paper theories, or analytical theories as we call them, can be formulated.

real sandpiles
and landscape
formation

Our ambitions extend beyond understanding the dynamics of real sandpiles. Nevertheless, experiments on sandpiles can be viewed as the first test of self-organized criticality. If the theory that large dynamic systems organize themselves to a critical state cannot even explain sandpiles, then what can it explain? Our abstract model grossly oversimplifies real sand, but we still hope that our experiments live up to our predictions. However, nature has no obligation to obey our ideas; our intuition could be entirely wrong. Theory has to be confronted eventually with real-world observations, so we study sandpiles and we ask, Do they or don't they self-organize to the critical state?

Long Island is blessed with miles of beautiful beaches, and Kurt Wiesenfeld was eager to do his own experiment. Soon after we came up with the sandpile idea, Kurt went to Smith Point Beach, ten miles south of the laboratory, and collected a small box of wet sand. He formed a steep pile of sand in the box, and let it relax until it came to rest. Instead of dropping sand on the pile, or tilting the box, he simply put the box on his windowsill so that the sun would slowly dry the pile. As the sand dried, the steep pile would become unstable, and there would be avalanches of sand falling off the pile to the bottom of the box and making the pile more shallow, possibly keeping the system at the critical state. As Kurt studied the sandpile, there indeed appeared to be avalanches of many different sizes.

Controlled experiments with sand test the robustness of our prediction of self-organized criticality (SOC). Following the publication of our theoretical sandpile model there was a spurt of worldwide experimental activity, including experiments on sand and other granular materials at the University of Chicago and at IBM, an experiment on rice in Oslo, Norway, and an experiment on mud slides in Hungary. The latter type of experiment may help us understand landscape formation. Sand slides onto roads in the Himalayas can be interpreted in terms of self-organized criticality. Sedimentary rock formation can be seen as evidence of avalanches that were formed on a geological time scale, indicating that landscape formation may be a self-organized critical process. The diversity of these experiments and observations underscores the resiliency of the phenomenon.

Real Sand

The experiments on sand turned out to be much more complicated and tedious than we had anticipated. Experiments must deal with length scales from as small as a grain of sand to thousands of times larger. The sandpiles must be very large to test the predicted power law behavior. In nature, where landscapes extend over thousands of miles, these various length ranges are readily available, but in real life we are restricted by limited laboratory space. Also, there is a limited amount of time available; one cannot wait for hundreds of years to amass a sufficiently large amount of data. On the computer, we had the luxury of studying billions of grains of sand and millions of avalanches. The distribution of avalanches is a power law, so large events are bound to occur; however, to have just one avalanche of size 1 million, one must wait for and monitor 1 million avalanches of size 1 (Figure 13). Experimentalists do not have that luxury.

The first experiment was performed by Sidney Nagel and Heinz Jaeger working with Leo Kadanoff at the University of Chicago. Kadanoff and his coworkers had been actively involved in much of chaos science, which had its heyday in the 1980s. Kadanoff has an eminently clear sense for good science; he taught me that good science is fun. There is always a lively atmosphere around him, and many animated discussions have taken place in his office and in the evenings under the influence of Kadanoff's single malt Scotch whiskey.

It was not surprising to me that Kadanoff and his colleagues were among the first to try to find a mathematical solution to the sand model that we had studied on the computer and to do the relevant experiments. Jaeger and Nagel partially filled a cylindrical drum with grains, and rotated the drum slowly, like a concrete blender. Turning the drum creates a sandpile at one side of the drum (Figure 14). The rotation makes the slope increase; now and then there are avalanches of falling sand, which reduce the slope. The reader can do his own experiment by tilting a bowl of sugar slowly, and observing the avalanches. The sand in the drum enters a stationary state with a well-defined average slope. However, the pile appears not to be critical in this stationary state. Indeed, there were very many small and intermediate-size avalanches with a distribution following the power law. However, once the avalanches reached a certain size, then inertial effects would kick in. Once a grain was

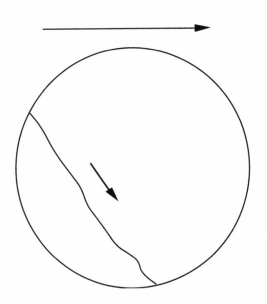

Figure 14.   Rotating drum experiment. The sandpile forms at one side of the drum as it is turning, and releases avalanches. This type of experiment was performed by the Chicago group, led by Heinrich Jaeger.

moving, it would gain momentum, and cause the avalanche to continue running until the whole pile came to rest at an angle that was a couple of degrees lower. Then the pile would start growing again because of the rotation. It would emit small and intermediate-size avalanches while building up its slope, until another enormous avalanche occurred. Thus, in its stationary state the pile exhibited an oscillatory motion by which the slope builds up and relaxes. This is not the critical behavior that we had predicted. The inertial effects responsible for this oscillatory behavior were not included in our simple model.

Glen Held and coworkers at IBM's research center in Yorktown Heights, New York, set up a different type of experiment, more in line with our initial suggestion. Their experimental setup is shown in Plate 2. They built a sandpile on top of a circular plate. This plate, with a diameter of a couple of inches, was placed on top of a precision scale. To make the experiment "authentic," they also collected the sand from Smith Point Beach near Brookhaven Laboratory. Sand was dropped at a very slow rate in the center of the plate through a slowly rotating glass tube. The sand formed a conical pile on top of the plate. The weight of the entire pile on the plate was recorded electronically, and the weight signal was sent to a PC for analysis at short time intervals. The computer calculated the mass of the avalanches of sand leaving the edges of the plate.

Held's team found behavior that was consistent with what Jaeger's team found. There was a range of avalanches with power law behavior; they found large avalanches causing oscillations of the slope of the pile, but they did not find intermediate avalanches. Their setup differed from the geometry that we suggested in one important aspect: only the amount of flow over the rim of the plate was recorded. The much more frequent internal avalanches, in which sand would move downward without leaving the plate, were not measured because they did not cause the weight of the pile to change.

We were encouraged by these very preliminary experiments because they revealed avalanches of many sizes. Nevertheless, some observers focused on the less than perfect agreement. John Horgan, a science writer at *Scientific American*, years later started a one-man crusade against complexity theory in general and self-organized criticality in particular. "Self-organized criticality does not even explain sandpiles," Mr. Horgan wrote gleefully, but completely out of context. While a good deal of skepticism is healthy, it would be better if science writers would let scientists themselves figure out what is right and what is wrong through the usual scientific process, which works pretty well in the long run. Usually, science writers go to the opposite extreme—they are too gullible, which is not as bad. I can assure the reader that my scientific colleagues can be relied upon to debunk what should be debunked.

Soon after these early experiments, Michael Bretz, Franco Nori, and their coworkers at the University of Michigan tried again with an elegant video technique. They performed two types of experiments. In one, they placed the sand in a Plexiglas box that was slowly rotated. The geometry of the experiment was that of an inclining ramp, similar to the Chicago experiment. They monitored the falling grains with a video camera, and sent the signal to a computer. By performing a digital image analysis of the pictures, they identified and measured all avalanches, including the internal ones that did not reach the edge of the pile. Bretz and Nori observed a power law distribution of avalanches. However, their system was small, and they had to halt the experiment when the sand stopped covering the bottom of the box; thus, the process could not continue indefinitely, as in the rotating cylinder. Bretz and Nori also performed an experiment with a sandpile onto which sand was dropped slowly, which was the geometry we had in mind. This experiment was also recorded by a video recorder (Plate 3), and found a power law distribution of avalanches with an exponent 2.13. These early experiments led to the inescapable conclusion that not everything in this world is SOC. Some of the piles are ticking periodically rather than flowing in bursts of all sizes.

Norwegian Rice Piles

The most careful experiment is a quite recent one performed by a group at the University of Oslo, Norway. Jens Feder and Torstein Joessang are the dynamic leaders of the Norwegian group, which is known for studies of fractal structures in porous media. In particular, they have made experimental and theoretical investigations on how liquids invade porous materials, which is of importance for the exploration and retrieval of oil in the North Sea and elsewhere.

Other investigators on this particular experiment were Vidar Frette, a graduate student, and Kim Christensen. Kim had already been working with us at Brookhaven on theoretical aspects of SOC and had played a prominent role in applying SOC to earthquakes, which is the subject of the next chapter. The final member of the team, Paul Meakin, formerly at DuPont research in the United States, is famous for large-scale simulations of growth of fractal structures. After the collapse of fundamental research at DuPont, following the general trend in industry in the United States, Meakin joined the Oslo group.

These scientists together created the ultimate sandpile experiment. One hopes this is a sign of things to come. Now that the multibillion dollar funding for the Texas superconducting super collider has vanished, experiments based more on thinking and imagination and less on the blind and mindless use of costly hardware, as had prevailed for thirty years, wouldn't be such a bad outcome. I suspect that more insight will come out of sandpile-type experiments than would ever have come out of the super collider, at a cost reduced by a factor of 10,000. But we will never know. Unfortunately the SSC was cancelled because of the general anti science attitude in the United States. None of the funding was transferred to other projects, but additional cuts were made everywhere. But since thoughts and sand are free, our research is more resilient.

Dr. Frette and coworkers chose to study grains of rice, not sand. In principle, it should not matter very much what kind of material is used. The details should not be important. The grains of rice have a convenient size that allows for a visual study of the motion of individual grains. The sandpiles with beach-type sand have problems because of the inertia of sand, which was not incorporated into our computer models.

The Norwegian group first went to the local supermarket to buy different types of rice. One type was almost spherical, and the experiments on that type had an outcome similar to the early experiments on sand. However, another type had long grains, which have more friction than sand and do not keep rolling. They are more likely to get stuck again once they start sliding.

The experiment was designed to be similar to our computer models exhibiting self-organized criticality, so it was important to monitor the bulk avalanches and not just the rice falling off the edges. The rice pile was confined to the space between two glass plates, through which the dynamics of the pile could be observed, either directly or with a video recorder. The rice was slowly fed into the gap at the upper corner by a seed machine at a slow rate of twenty grains per minute. Experiments were performed at various spacings between the plates and at various slow feeding rates. Experiments were also performed on many different system sizes, ranging from a few centimeters to several meters. Each experiment lasted forty-two hours, so the dedicated participants had to take turns staying overnight to supervise the experiment. The long runs were important in order to get good statistics, particularly for the very few large avalanches, and the large system size was important in order to have a wide range of avalanche sizes. In total, the experiment with the various sizes and types of rice ran for over a year!

Motion of the grains was monitored with a CCD video camera with 2000 × 500 pixels covering the active area. Frames were then taken every fifteen seconds, and the digitized signal was sent to a computer, identifying the positions of all the rice grains. The pile grew until it reached a stationary state. Once the stationary state was reached, the camera and the computer started monitoring the motion of rice grains. Figure 15 shows a propagating avalanche in the stationary state. The profiles of the rice pile at two consecutive measurements are shown. The gray area shows the rice that was present at the first measurement, and not at the second, i.e., the amount of rice that had fallen. Conversely, the black areas show where the rice went. Those areas were not filled at the first measurement, only after the second. Thus, an avalanche had occurred in the fifteen-second interval between the two measurements. The size of an avalanche was defined as the total amount of downward motion of grains between two successive frames, that is the number of grains falling weighted by the distance they fell. The size of the avalanche measured this way is equal to the energy lost, or dissipated into heat.

In the stationary state, the rice grains get stuck in intricate arrangements, where they lock into each other, allowing for steep slopes, even with overhangs (Plate 4). An analysis of the surface profile shows that it is a fractal structure just like the coast of Norway, with bumps and other features of all sizes. The

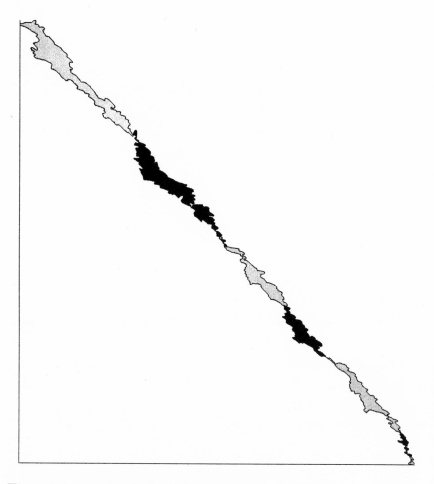

Figure 15. Avalanche in the Norwegian rice pile experiment. During a fifteen-second interval, rice left the gray areas and ended up in the black areas (Frette et al., 1995).

fragility of the critical state as compared with a flat bowl of rice is evident from the figure.

Figure 16 shows a sequence of avalanches that occurred in a period of 350 minutes during one run. On the basis of such measurements, one can count the number of avalanches of each size. For the long grain rice the distribution of avalanches is a power law, indicative of SOC behavior. The distribution was measured for different sizes of piles (Figure 17). The larger the pile, the larger the avalanche. The same scaling behavior was observed for avalanches ranging in

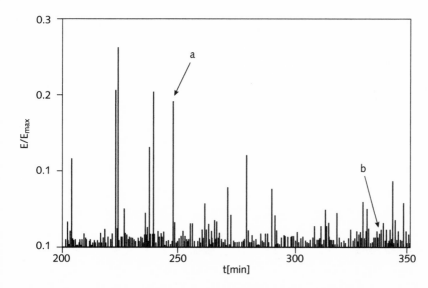

Figure 16. Avalanches measured during a period of 350 minutes. The heights of the lines are proportional to the size of the avalanche.

size from a few grains to several thousand grains. Frette and coworkers showed that the curves for different sizes of systems followed a systematic behavior, known as "finite size scaling," unique to critical systems. Thus, SOC can indeed be observed in the laboratory sandpiles, if one has persistence and patience.

By coloring a few grains, the experimenters were able to trace the motion of the individual grains. This turned out to be surprisingly complicated. The sliding grains were not confined to the surface; the grains made complicated excursions of long duration through the pile. No grains would stay forever in the pile. They all eventually would leave, but some grains remained in the pile for an extremely long time. This behavior is not understood at all, but it does not affect the SOC behavior, as evidenced by the measured power law. It would be interesting if the duration of grains conformed to another power law.

Experimentalists might wish to have avalanches spanning an even larger range of magnitude in experiments of longer duration. However, no laboratory experimentalist has the infinite patience nature has, and no laboratory has the space nature has, so there are limits on the systems that can be studied. Observations of real phenomena, such as the distribution of earthquake magnitudes,

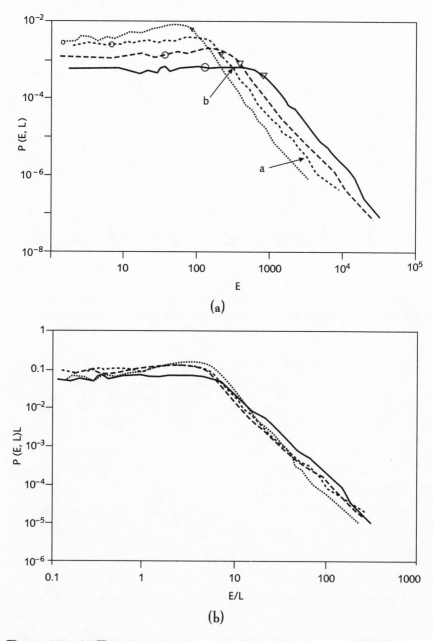

Figure 17. (a) Distribution of the size E of avalanches in rice piles of different lengths L. The various curves are for various sizes of rice piles. By systematically shifting the curves, they can be brought to cover each other (b). This property is known as finite size scaling, and implies criticality (Frette et al., 1995).

might show scaling behavior, i.e. power laws, over much wider ranges than the short-term experiments in the lab. After all, it took billions of years for the morphology of the earth to reach its present state. On the other hand, laboratory experiments allow study under systematically varying conditions, whereas nature represents only a single experiment. This is the problem that one generally encounters when studying emergent phenomena such as large avalanches: the experiment must contain everything from the shortest length scale of the microscopic entities to the largest where the emergent phenomena occur. In contrast, the "reductionist" scientist sees a need to study things at only the smallest scale.

Nevertheless, the Norwegian rice experiments show conclusively that SOC occurs in piles of granular material within the limits defined by the laboratory conditions.

Vicsek's Landslide Experiment: The Origin of Fractals

Tamas Vicsek is a Hungarian physicist who has devoted most of his career to studying fractal phenomena. Together with Fareydoon Family at Emory University in Atlanta, he developed a general formalism for describing growth of surfaces by random deposition of material. The theory, known as the Family–Vicsek scaling, is widely used both by experimentalists and theorists. Recently, Vicsek has constructed a fascinating model for self-organization of a flock of birds. He showed that it was possible for the birds to fly in formation in the same direction without a leader. The individual birds would simply follow their neighbors. The flock migration is a collective effect, as is SOC.

In collaboration with colleagues at the Eotvos University in Budapest, E. Somfai and A. Czirok, Vicsek did an experiment that not only confirmed the evolution of a sandpile to the critical state, but also threw light on the mechanisms for landscape formation in nature. Why do landscapes look the way they do? They decided to build their own mini-landscape, subjected to erosion by water. This type of laboratory experiment may be an interesting contribution to geomorphology, the science of how real geological structures are formed.

A granular pile was erected by slowly pouring a mixture of silica and pot soil onto a table. The initial "landscape" had the shape of a ridge. The ridge was watered by commercial sprayers modified to suit the experiment (note again, a low-budget experiment). As the water penetrated the granular pile, parts of the pile became saturated, and these wet parts slid down the surface, like avalanches or mud slides.

The purpose of the experiment was to gather information on the distribution of the sizes of the landslides in this micro-model of landscape formation by water erosion. This was done by video recording the changes in the profile of the ridge, just as Frette and coworkers did for the rice pile. The information was fed to a computer for analysis.

Since each experiment eventually caused a complete breakdown of the pile, the experiment had to be repeated many times to get a sufficiently large number of avalanches. In principle, to represent real landscape formation, the watering-down should be balanced by some kind of landscape upheaval. In all, Vicsek and coworkers performed nine independent erosion experiments with between ten and thirty mudslides in each experiment. All the data were combined to form a single histogram of landslide sizes, which exhibited a power law shape with an exponent near 1, indicating self-organized criticality.

The experimenters measured many other properties of the landscapes formed by the erosion process. The distribution of velocities of the landslides is another power law. Most importantly, they measured the geometrical properties of the resulting contours of the landscape. They found that it is a fractal with features at all length scales! Thus, Vicsek's group had demonstrated in a real experiment that fractals can be generated by a self-organized critical process, precisely as predicted from the sandpile simulations and as found also by the Norwegian group.

Mandelbrot, who coined the term *fractal,* rarely addressed the all-important question of the dynamical origin of fractals in nature, but restricted himself to the geometrical characterization of fractal phenomena. The Hungarian experiment showed directly that fractals can emerge as the result of intermittent punctuations, or avalanches, carving out features of all length scales.

Thus it is a very tempting suggestion that fractals can be viewed as snapshots of SOC dynamical processes! In real life, where time scales are much longer than in the laboratory, landscapes may appear static, so it may not be clear that we are dealing with an evolving dynamical process. In the past, geophysicists have fallen into this trap when dealing, for instance, with earthquakes as a phenomenon occurring in a preexisting fault structure. The chicken (geometric fractal structure of the network of faults, or the morphology of landscapes) and the egg (earthquakes, landslides) were treated as two entirely different phenomena. The geophysicists did not realize that the earthquakes and the fault structures could be two sides of the same coin, different manifestations of one unique underlying critical dynamical process.

Himalayan Sandpiles

Do sand slides in nature obey the power laws indicative of SOC that were observed in the laboratory under controlled circumstances? To shed some light on this, David Noever of the NASA George C. Marshall Space Flight Center in Alabama has investigated sand slides in the Himalayas. Noever examined data from two road-engineering projects. On two mountain roads in Nepal, the six-kilometer Mussoori-Tehrie road and the two-kilometer stretch on the recently completed Mussoori bypass, avalanches were cleared off the road. The smallest landslides had a volume of $1/1000$ cubic meters, which is about a shovelful. The largest avalanches were $10,000,000$ cubic meters, so the landslide volumes spanned a colossal range of eleven orders of magnitude, compared with the two or three orders of magnitude covered by the laboratory experiments.

In contrast to the early sandpile experiments, there were events of all sizes. The distribution of avalanches follows a power law over about six orders of magnitude. The power law was not obeyed for avalanches smaller than one cubic meter. I suspect that this is simply because not all avalanches involving a few shovelfuls were recorded, just as not all small earthquakes are. (See Figure 2 for a similar effect for small earthquakes.) Also, the small sand slides may have been removed by cars and yaks traveling along the roads. In any case there was scaling extending over an enormous range. Noever notes that the avalanches originate from a steep "supercritical" state that erodes and produces avalanches. He points out that one obvious laboratory setup "would be systematically drying or vibrating an overly steep pile of wet sand." That was essentially the type of experiment that Kurt performed in an uncontrolled way on his office windowsill in 1987.

Sediment Deposition

Rocks formed by sediment deposition form a layered structure. One process for the formation of the layers works as follows. First, by various transport processes, sediment is deposited at the edge of the continental shelf and along the continental slopes. The slope eventually becomes unstable, causing avalanche-like events known as *slumps*. The slump creates a region of mud, which flows along the sea bottom. Eventually the mud current slows down when it reaches the relatively flat basin plain, at which point the sediment it has carried finally settles down. Deposits produced this way are called *turbidites*. Turbidite events occur on time scales ranging from minutes to days, whereas the time between deposition events in any location is thought to be on the order of years to thousands of years. We are dealing with an intermittent, punctuated equilibrium phenomenon. By studying the thickness of layers, ranging from centimeters to several meters, one can estimate the distribution of avalanches causing the sedimentation.

Some of the experiments on sandpiles did not exhibit SOC, presumably because of inertia effects, in contrast to our model, which did not include the inertia, or momentum, of the sand grains. This observation is intriguing and relevant to the interpretation of turbidite deposition. Since the slumping occurs in the ocean, the water may be sufficient to damp the motion.

Daniel Rothman of MIT and his collaborators John Grotzinger and Peter Flemings have carried out a detailed study of turbidite deposits. Turbidites can be observed at the Kingston Peak formation along the Amargosa River near the southern end of Death Valley, California (Figure 18). The turbidites were formed approximately 100 million years ago. The sample that Rothman's team studied was obtained by drilling a hole several hundred meters deep and recovering the sediments from that hole. They counted how many layers exceeded a certain thickness, and made the usual log-log histogram (Figure 19). Indeed, there is a power law distribution of layer thicknesses, as the theory of SOC predicts.

(a)

(b)

Figure 18.   Photographs of turbidites in the Kingston Peak Formation. Daniel Rothman, John P. Grotzinger, and Peter Flemings. Note the layered structure, spanning a wide range of thicknesses (a). The penny illustrates the scale (b).

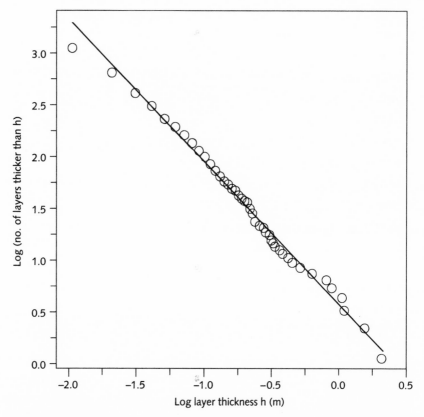

Figure 19.   Number of turbidite layers thicker than h as a function of
the logarithm of the layer thickness h for 1,235 turbidites observed in
the Kingston peak formation. The straight line has a slope of 1.39, indicating a power law distribution of the thicknesses of the layers, and thus
for the distribution of avalanches which formed them (Rothman et al.,
1994).

Geomorphology:
Landscapes Out of Balance

Landscapes are prime examples of complex systems. Simple systems do not
vary much from one place to another. Landscapes are different. We can look
around and orient ourselves by studying the landscape precisely because every
place is different from every other place. Complexity involves surprises. Every
time we turn a corner, we see something new. What are the general principles

governing the formation of landscapes? So far, there has been no general framework for discussing and describing landscape formation.

It puzzles me that geophysicists often show little interest in the underlying principles of their science. Perhaps they take it for granted that the earth is so complicated and messy that no general principles apply, and that no general theory (in the physicist's sense) can exist. There are outstanding exceptions, however. Donald Turcotte of Cornell University has been involved in discovering the general mechanisms governing geophysics for a number of years. In particular, he has performed extended analysis of many fractal phenomena and constructed simple mathematical models reproducing some general features in geology and geophysics.

Another exception is Andrea Rinaldo of the University of Padova. His university may be considered the cradle of modern science. In the fifteenth century the idea of studying the human body by observing and describing, rather than by unsubstantiated philosophical arguments, originated at Padova.

Rinaldo is a hydrologist. He studies the flow of water on earth—in the ground, the oceans, and the atmosphere—and the interactions between water and vegetation. He has been particularly interested in the complicated dynamics of the flow of water from the Adriatic Sea back and forth into the lagoons of Venice. In the best tradition of the University of Padova, Rinaldo wants to identify some general principles for the formation of landscapes. Together with his colleagues Riccardo Rigon, also of Padova, and Ignacio Rodrigues-Iturbe, a colorful and outspoken geophysicist from Venezuela, he initiated a theoretical study of the formation of river networks and the effects of the rivers on landscapes. Small rivers, starting essentially everywhere, join each other to form larger rivers, which merge to form even larger rivers, and so on until the largest rivers run into the oceans.

It is known that the branching structure of rivers follows a simple power law known as Horton's law. Horton defined the order of river segments as the number of links to other segments that has to be passed before the river reaches the ocean. Horton's law states that the number of segments of each order increases as a power law in the order. This hierarchical structure indicates that river networks are fractal, just as the hierarchical structure of fjords along Norway's coast indicates that the coast is fractal. Another empirical law says that the length L of a river scales with the area A that is drained by that river as

$$L = 1.4A^{0.6}.$$

Could it be that these and other power laws for river networks are indicative of SOC?

In sandpile models, the criticality comes about from a combination of two processes: energy is supplied by adding sand or tilting the pile, and energy is dissipated by toppling of the grains of sand. Rinaldo's group speculated that landscape formation occurs by a similar process, in which energy is supplied by an uplifting process (by plate tectonic or some other geological process) and dissipated through erosion by wind and water.

In Rinaldo's model, erosion takes place if the stress on a riverbank from the water flow exceeds a critical value. The stress at a given point depends on the flow of water through that point, and the slope s of the landscape. The flow of the water is proportional to the area A that is drained by the river branch, assuming that the rain falls at the same rate everywhere. The formula for the stress was taken to be

$$\text{stress} = \sqrt{As^3}$$

(although the exact expression is not important).

The simulation is quite simple: Starting from a given landscape with a river network, the stress is calculated everywhere using the formula above. The sites where the stress exceeds the critical value are identified. Erosion is simulated by removing one unit of material at each of those sites. After the erosion takes place, a new landscape, with a new network of rivers, has emerged, and the process is repeated. The river pattern is constructed from the resulting contours of the landscape by having the water always running in the direction of steepest descent from any point. The erosion is combined with a general uplifting that uniformly increases the slope s of the landscape everywhere. It would be interesting to do real laboratory experiments of the type that Vicsek did, in which the washing down of the sandpile is combined with uplifting, for instance a gradual tilting of the pile.

The landscape settles into a stationary state, with a fractal network of rivers traversing a fractal landscape. Figure 20 shows a snapshot of the river network. Many aspects of the computed river network are in agreement with empirical observations, such as Horton's law and the law for the drainage area for a river of a given length. The power laws show that the stationary state is critical. Plate 5 shows the corresponding landscape that was generated by the process.

Rinaldo's computer simulations of landscape formation represent a new and refreshing way of looking at geophysics. Instead of simply describing all

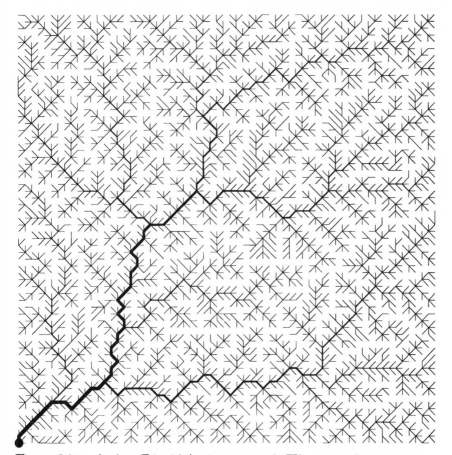

Figure 20.   Andrea Rinaldo's river network. The network was generated by a computer calculation following a simple rule for erosion. The network has statistical properties similar to those of real river networks.

geophysical features by a simple cataloging process, or "stamp collection," the simulations reveal the general mechanisms. Observing details may be entertaining and fascinating, but we learn from the generalities.

Rinaldo concludes that the fractal structure of river networks on the surface of the earth is a manifestation that the crust of the earth has self-organized into a critical state, forming landscapes "out of balance." No other dynamic mechanism for the formation of fractals in geophysics has been proposed. The variability of landscapes can be viewed as an SOC phenomenon. Landscapes are snapshots of a dynamic critical process.

It is particularly rewarding to visit Rinaldo and his group. Our meetings take place at the Instituto Veneto Di Scienze, Lettere Ed Arti, an impressive classical building in the heart of Venice, within ten minutes walk of both the Rialto Bridge and Marcus Square. It forms a remarkable contrast to the barracks in which we work at Brookhaven Laboratory. The environment has a stimulating effect on the lively discussions about nature at work.

In the final analysis, it is these applications of the ideas of self-organized criticality to real features in our world that make our theoretical effort worthwhile. Self-organized criticality is not just an esoteric mathematical computer game; it represents an explanation of everyday objects in nature. More examples linking self-organized criticality with the dynamics of nature follow.

earthquakes, starquakes, and solar flares

Earthquakes may be the cleanest and most direct example of a self-organized critical phenomenon in nature. Most of the time the crust of the earth is at rest, in periods of stasis. Every now and then the apparent tranquillity is interrupted by bursts of intermittent, sometimes violent, activity. There are a few very large earthquakes and many more smaller earthquakes. The small earthquakes do not affect us at all, so scientific efforts have been directed toward trying to predict the few large catastrophic ones. Scientists have taken a very direct approach, formulating individual theories, or explanations, for individual earthquakes or earthquake zones; there has not been much effort directed toward a general understanding of the earthquake phenomenon. The geophysics community is very conservative. For instance, the theory of plate tectonics as a general explanation for the shifting of crustal plates that creates earthquakes was put forward in *The Origin of Continents and Oceans* by the German meteorologist Alfred Wegener in 1912, but not even found worthy of discussion until the late 1960s. Among its obvious appealing features, it explains the similar shape and geological composition of the west coast of Africa and the east coast of South America.

Don't get me wrong. I have the deepest respect for the type of science where you put on your rubber boots and walk out into the field to collect data about specific events. Such science provides the bread and butter for all scientific enterprise. I just wish there was a more open-minded attitude toward attempts to view things in a larger context.

I once raised this issue among a group, not of geophysicists, but of cosmologists at a high table dinner at the Churchill College in Cambridge. "Why is it that you guys are so conservative in your views, in the face of the almost complete lack of understanding of what is going on in your field?" I asked. The answer was as simple as it was surprising. "If we don't accept some common picture of the universe, however unsupported by the facts, there would be nothing to bind us together as a scientific community. Since it is unlikely that any picture that we use will be falsified in our lifetime, one theory is as good as any other." The explanation was social, not scientific.

Explanations for earthquakes typically relate the earthquakes to specific ruptures of specific faults or fault segments. This might be reasonable, but then, of course, one has to explain the fault pattern independently. Analogously, our sand man may correctly conclude that the origin of sand slides is toppling sand, but that does not provide any insight into the properties of large slides. The fact that earthquakes are caused by ruptures at or near faults does not in itself explain the remarkable Gutenberg–Richter law.

Scientists are poor at making earthquake predictions, and not for lack of effort. All kinds of phenomena in nature have been viewed as precursors of large earthquakes, such as the behavior of animals, the variations in the ground water level, and the occurrence of minor earthquakes. The latter approach, trying to recognize earthquake patterns preceding major quakes, seems, at least in principle, plausible. However, there has been no success. In particular, there have been claims that earthquakes are periodic at some locations, but the statistics were never based on more than two to four intervals. Notably, it appeared that in the Park Field earthquake region in California there was a periodicity of approximately 20 years. Some years ago a major and expensive project was set up to study the next earthquake. The last event in that area took place in the 1950s, and the scientists are still waiting! The earth-quake predictors have had much less success than their meteorologist col-leagues. "Only fools, charlatans, and liars predict earthquakes," Richter (fa-ther of the Gutenberg–Richter law, and the Richter scale for earthquake magnitudes) once said. The phenomenon is surrounded by much folklore. Because of the poor statistics of the very few large quakes, one can say just about anything about earthquakes without being subjected to possible falsification. The predictions will not be challenged within the lifetime of the person making the prediction.

Indeed, after an earthquake one can report what happened in some de-tail. One can identify the fault that was responsible and pinpoint the epicen-ter. This information might convince scientists working on earthquakes that one should be able to predict large events. "With a little more funding" one might become successful. However, our experience with sandpile modeling tells us that things do not generally work out that way. Because we can explain with utmost precision what has happened does not mean that we are able to predict what will happen.

It seems reasonable to take some time to acquire a general understanding of earthquakes before jumping into predicting specific events. This chapter discusses the extensive work that has been performed during the last few years, supporting the view that earthquakes are an SOC phenomenon. The Guten-berg–Richter law—discovered long before anybody thought about land-scape self-organization—epitomizes what SOC is all about. The distribution of earthquake magnitudes is a power law, ranging from the smallest measur-able earthquake, whose size is like a truck passing by, to the largest devastating quakes killing hundreds of thousands of people. I cannot imagine a theory of earthquakes that does not explain the Gutenberg–Richter law.

The Gutenberg–Richter law (Figure 2) is a statistical scaling law—it states how many earthquakes there are of one size compared with how many there are of some other size. It does not say anything about a specific earth-quake. The law is an empirical law—it stems from direct measurements and has not previously been connected with general principles in physics.

One might think that there is something special about the *largest* events on the curve, of magnitude 9 or so for a worldwide catalogue. It appears that there must be some particular physics on the scale that prevents larger quakes from taking place. This is probably an illusion. The largest events merely represent the largest magnitude that we typically can expect in a human lifetime. Even if the Gutenberg–Richter law extends beyond earthquakes of magnitude 10, we may not have had the opportunity to observe even a single one. A superhuman living for a million years might well have observed a few earthquakes of magnitude 12, involving, for example, most of the earthquake zone ranging from Alaska to the southern tip of South America. To this superhuman, earthquakes of magnitude 9 might appear uninteresting. Similarly, a mouse living only for a year or so, might find an earthquake of magnitude 6 terribly interesting, since this is the largest it can expect to experience in its lifetime. Unfortunately, it is not yet possible to check by geological observations whether or not there have been earthquakes of magnitude, say, 10 in the last 10,000 years.

The scaling law says that there can be nothing special about earthquakes of magnitude 8 or 9 because there is nothing special about a human lifetime of 100 years or so (the average time interval for such events) in a geophysical context, in which the time scale for tectonic plate motion is hundreds of millions of years. That is not necessarily a bad situation; since the physics is the same on all scales, one might acquire insight into earthquakes of magnitude 8 or 9 by studying the much more abundant quakes of magnitude 5 or 6, the statistics of which are more available. It is pointless to hang around for dozens of years to get better data on large earthquakes.

Self-Organization of Earthquakes

I first heard about the Gutenberg–Richter law in 1988 during a Gordon conference on fractals, soon after our discovery of SOC. Gordon conferences are informal, private conferences where scientists in many different areas can present and discuss their most recent results. The Gordon conference is a magnificent institution that has served science very well. They take place in the summer at a few small colleges near the beautiful lakes, forests, and mountains of New Hampshire, and offer an opportunity to combine scientific discussions with a variety of recreational activities. As we have seen before, the environment plays a large role in human scientific creativity.

Scientists in general, and physicists in particular, are quite enterprising when it comes to selecting sites for their communal activities. My career as a physicist has allowed me to visit some of the most fantastic places on Earth, such as Aspen, Colorado (the Aspen Center of Physics), Santa Barbara (the Institute of Theoretical Physics), New Hampshire, Venice, the Great Wall of China, Moscow (the Landau Institute), Santa Fe (the Santa Fe Institute), and the Alps (the Physics Institute in Les Houches, near Chamonix). You don't get rich from doing physics, but you do get an opportunity to go to all the places the rich would go to if they had the time.

The Gordon conference was on fractal structures in nature. It was particularly stimulating because it brought together scientists from many different fields. More typical scientific conferences deal with narrow esoteric subjects about which all the participants are experts. One of the speakers of this conference was Yakov Kagan of UCLA, who addressed the importance of scale-free behavior of earthquakes and earthquake zones. He pointed out that faults form fractal patterns, and presented worldwide earthquake data showing power law behavior of earthquake magnitudes over seven decades. This was the first time I had heard about the Gutenberg–Richter law.

Kagan gave a sharp rebuttal to much of the folklore surrounding the earthquake business, such as "characteristic earthquake" sizes. I had never been involved professionally in geophysics and knew little about the subject. Nevertheless, I was fascinated by his talk. Were earthquakes like the sand slides in our sandpile model? Tectonic plate motion, providing the energy for the earthquakes, would correspond to tilting the sandpiles in the ramp version of the model. The ruptures would correspond to toppling grains. Just as the increased force on the grains from the slow tilt would necessarily sooner or later cause the sand to topple somewhere, the slowly increasing pressure from the tectonic plates grinding into one another eventually must cause rupture somewhere. Just as toppling grains can affect one another in a domino process, one rupture can lead to another by the transfer of force, and sometimes lead to a large chain reaction representing a large earthquake. In a larger perspective, one might think of the plate motion as the source of "landscape upheaval" and the earthquake as the "erosion," whose combined effects organize the crust of the earth to the critical state, using Rinaldo's "landscape out of balance" picture.

I returned to my laboratory and, together with Chao Tang, did some further computer simulations of the sand model. We studied the continuous, deterministic version in which the sandpile is slowly tilted, which is the version with real variables Z in Chapter 3.

What we had in mind was a block-spring picture of earthquake generation (Figure 21), in which the fault is represented by a two-dimensional array of blocks in contact with a rough surface. In the real world one cannot localize the earthquakes to single preexisting faults. The Gutenberg–Richter law concerns the statistics of earthquakes over an extended region like California. Of course, we cannot construct a realistic computer model of California and follow its evolution through hundreds of millions of years, as we would like to do. In the block-spring model, the blocks are connected to a constantly moving plate by leaf springs. The leaf springs represent the pressure on the material near the fault due to the tectonic plate motion. The blocks are also connected with each other by coil springs. Each element sticks to the surface when the sum of the spring forces is less than a threshold. The leaf springs exert a constantly increasing force on all the blocks. When the force on a particular block becomes larger than the threshold, the block slips instanta-

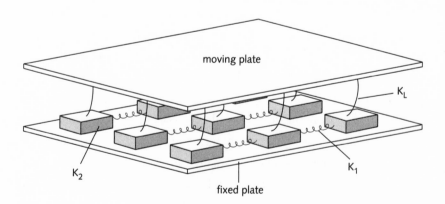

Figure 21.   Block-spring model of earthquake fault. The blocks are connected with a slowly moving rod by leaf springs. They are also connected with each other by springs. Parameters K_1, K_2, and K_L specify the strengths of the springs. The blocks are moving on a rough surface. A blocks slide when the force on it exceeds a critical value.

neously in the direction of the moving rod. Because of the coil springs, this in-creases the force on the four neighboring springs, and might cause the force on one or more of those blocks to exceed the critical value, so that they, too, would slip. This could lead to the chain reaction representing the earthquake. This type of model had been introduced years before, in 1967, by Burridge and Knopoff at UCLA.

It was payoff time for our work on the rotating pendulums. The arith-metic of the block-spring model was very much the same as for the coupled pendulums. Pulling the blocks by the leaf spring was like slowly winding up all the pendulums simultaneously, until one of them would make a rotation, initializing an avalanche. The slip of a block corresponds to the rotation of a pendulum. In turn, we knew that the rotation of a pendulum is equivalent to the toppling of a grain of sand in the sand model. Thus, the three models are mathematically identical; if you have studied one, you have studied them all! Indeed, it was at that point we found that the continuous, slowly driven deterministic sandpile model provided the same power law as the initial sto-chastic version, driven by adding sand randomly. Thus, the Gutenberg–Richter law is the fingerprint that the crust of the earth has self-organized to the critical state.

Soon after, other groups independently discovered that earthquakes can be thought of as a SOC phenomenon. Didier and Anne Sornette, a married couple at the University of Nice, presented their results in a short article in *Europhysics Letters;* they pointed out the analogy between sandpile models and block-spring models. Didier Sornette may be the most imaginative of all geo-physicists—maybe even too imaginative, if that's possible. Every six months he comes up with another general observation or theory of some geophysical phenomenon. His batting average of being right is rather low, but in science that doesn't matter, as long as just once in your lifetime you say something important and correct. Keisuke Ito and Mitsuhiro Matsusaki of Japan published a much more detailed account in the *Journal of Geophysical Re-search.* These authors also studied the possible origin of aftershocks, which also were known to follow a power law distribution known as Omori's law. Amaz-ingly, all three groups chose essentially the same title, "Earthquakes as a Self-organized Critical Phenomenon."

A fourth group, Jean Carlson and Jim Langer of the Institute for Theoretical Physics, Santa Barbara, made much more laborious calculations on a
more detailed model in which the blocks did not slip instantaneously to their
new positions, as represented by the toppling of a grain of sand in the sand
model, following Newton's law. They kept the inertia of the blocks, in contrast to the sandpile versions. This type of calculation is very slow, so only
small systems can be studied. It was precisely to avoid such calculations that
we introduced the simpler sand models instead of the messy rotating pendulums, which supposedly would behave in the same way. Another justification
for the simpler sand model is that we really don't know the forces, including
the friction, to insert into the block-spring model, so the model is not realistic
under any circumstances. Carlson and Langer found a power law for small
earthquakes, and they found more or less periodic huge earthquakes, a distribution not found for real earthquakes. Their simulation gave a much better
description of the early sandpile experiments performed by the Chicago
group, where inertial effects take over and prevent intermediate size
avalanches. Recall that, in contrast, the Norwegian group had reduced inertial effects by using long sticky grains of rice.

We were ambitious, and sent an account of our earthquake ideas to the
world's most prestigious journals, first to *Nature* and then to *Science*. Our article
was rejected by both journals, by geophysicists who did not understand what
it was all about. The idea of having a general theory of the phenomena of
earthquakes was unacceptable. However, the referees should be given credit
for revealing their identity, which is not required in the normally anonymous
refereeing process. To appreciate the pain and annoyance that one might feel
because of such a decision, it should be pointed out that essentially anything
can be published, no matter how insignificant—even in *Nature*. Most published material sinks like a rock and never surfaces again. It is precisely when
you have something potentially new and interesting that you get in trouble.
Ironically, dozens of articles applying our ideas to various natural phenomena have since appeared with great regularity in those same journals.

Soon after, I presented our ideas at a conference on earthquakes in Mon-
terey, California, a place with a spectacular view of the surf of the Pacific
Ocean. I couldn't help noting in my talk that our article had been rejected for

publication in *Nature* by Professor X who is sitting to the left, and for *Science* by Dr. Y who is sitting to the right. Both flushed. But at least everyone became aware of our ideas at that point. At the same conference, Jim Langer presented calculations on the more detailed Carlson-Langer one-dimensional block-spring models.

Eventually, our article was published in the *Journal of Geophysical Research* by its editor, Albert Tarantola, who took the matter in his own hands and published the article despite its rejection by his referees. By 1995 there were more than 100 articles in the literature supporting the view of earthquakes as an SOC phenomenon.

Our model was immensely oversimplified and wrong in one respect. Our original sand model was *conservative*, that is, all the sand that topples ends up at the neighboring sites. There is no sand lost in the process. That is quite reasonable for sandpiles. For earthquakes, on the other hand, a careful analysis of the block-spring model shows that there is no reason for conservation of forces. The amount of force that is transmitted to the neighbors may be less than the release of force on the sliding block. As soon as the condition of conservation was relaxed in the sand model, by letting not one grain of sand arrive at the neighbor sites, but, say only 0.9 grains, the Gutenberg–Richter law would be obeyed only up to a cutoff magnitude that would depend on the degree of conservation. There would be only small earthquakes. The block-spring model would not be critical!

A Misprint Leads to Progress

The solution to this problem was found by accident. In 1990 Kan Chen and I wrote an extended version of our earthquake article for a book, *Fractals in the Earth Sciences*, edited by Christopher Barton of the U.S. Geological Survey. Kan Chen was a research associate working with the condensed matter theory group, coming to us from Ohio State University. We had made extensive calculations on the continuous version of the sandpile, where all the heights are raised uniformly until there is an instability somewhere.

Barton had for some time been excited about the appearance of fractals everywhere in geophysics, and decided to edit a book on the subject, with chapters written by scientists working on fractals. Chris immediately saw the possibility of SOC being the underlying dynamic mechanism for a variety of geophysical phenomena, and asked me to write a chapter. Unfortunately, there was a minor misprint in the preprint of that article that we circulated to colleagues. The following discussion necessarily deals with some technical issues.

Let us recall from the discussion of the sand model that when the height, representing the force f acting on a particular part of the crust of the earth, reaches $f = 4$, it relaxes to $f - 4$, while transmitting one force unit to each of its four neighbor blocks. Instead, we wrote that f goes to 0. For the first toppling in an avalanche this is no problem, since f of the toppling site is exactly 4. However, for some of the subsequent toppling events f is greater than 4, so the relaxation is greater than 4, and only 4 units of force are transmitted. Thus, there is a net loss of force in the process if f is reset to 0.

In Oslo, Hans Jakob Feder, together with his father Jens Feder, decided to test the SOC earthquake theories by pulling a sheet of sandpaper across a carpet. The motion was not smooth, but jerky. They measured a power-law distribution of the sizes of the slip events. Hans was a high school student in Oslo, Norway at that time.

The Feders also decided to simulate earthquakes using our instructions in the preprint. Indeed, they reproduced the Gutenberg–Richter law, but with other exponents than the ones we predicted. Jens Feder called me, and the misprint was discovered. Inadvertently, they had studied a model that had *no conservation* of force, but nevertheless exhibited SOC. This was of great importance, since at that time there was a growing suspicion among scientists working on dynamic phase transitions, such as Geoffrey Grinstein at IBM and Mehran Kardar at MIT, that SOC occurred only if the system was "tuned" to be conservative, indicating that one would not in general observe criticality in nature. I had great difficulties rebutting those claims at that time. The Feders published their results in *Physical Review Letters*.

I decided to invite Hans Jakob Feder to Brookhaven in the summer of 1991. At that time I had two very resourceful research associates working with me, Kim Christensen, who later became involved in the Norwegian rice pile experiment, and Zeev Olami, a postdoctoral fellow from Israel. Kim was formally a graduate student of the University of Aarhus in Denmark, so the project was to be included in his thesis. In an earlier work, performed while an undergraduate student in Aarhus, Denmark, he showed that our analysis of $1/f$ noise in the original sandpile article was not fully correct. Fortunately, we have since been able to recover from that fiasco in a joint project by showing that for a large class of models, $1/f$ noise does indeed emerge in the SOC state. One could not imagine a more diverse pair of scientists. Kim works carefully, logically, and systematically; Zeev is intuitive, undisciplined, and full of ideas. This was an ideal collaboration, with Kim keeping Zeev honest by flushing out the worst ideas.

Zeev, Kim, and Hans Jakob started with the block-spring picture (Figure 21), and transformed it into a mathematical "sandpile"-like algorithm: Each block is subjected, as usual, to a constantly increasing force from the moving rod, and a force from the neighbor blocks. Whenever the force on any block exceeds the critical value $f = 4$, the force on that block is reduced to 0, while a fraction α of that force is transferred to each of its four neighbors. In the special case that the fraction α is $1/4$, the model reduces to the deterministic conservative version of the original sand model. When α is less than $1/4$ the model is nonconservative.

It cannot be emphasized enough that the setup in Figure 21 does not really represent how earthquakes work. It is our spherical cow. The earthquake cannot be localized to individual, preexisting faults; it is a three-dimensional distributed phenomenon. The Gutenberg–Richter law is not a property of a fault, but a property of the entire crust, or at the very least a large geographical area. Ideally, we would like to have the fractal systems of faults be created by the earthquake dynamics itself in the model, simulating the entire geological process that formed the crust and eventually carried it to the critical state. The model is merely intended to show that such behavior is indeed within the realm of the possible.

Hans Jakob, Kim, and Zeev studied the model on the computer. Indeed, they found earthquakes of all sizes following the Gutenberg–Richter law! (Figure 22). What was particularly interesting about this result was (1) the model was derived from a careful analysis of the original Burridge-Knopoff block-spring model, which was already well known and accepted in the community (they did not have to pull some new "ad hoc" physics out of the hat);

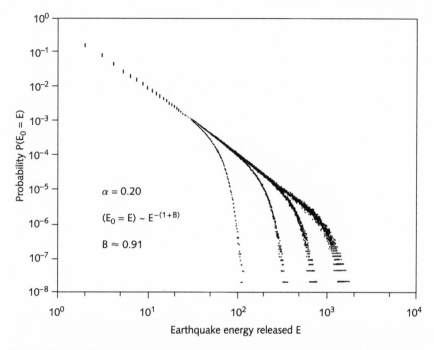

Figure 22. Gutenberg–Richter law for synthetic earthquake catalogs originating from the block-spring model studied by Olami et al. (1992). The various curves correspond to different system sizes.

and (2) the model required no tuning in order to be critical. The power law was valid for a wide range of values of the parameter α. They could even include various types of randomness in their model without destroying the criticality.

The various curves in the figure correspond to various numbers of blocks. When the number of blocks in the system increases, the power law extends to larger events in a systematic way known as finite size scaling, which only critical systems obey. Conversely, if the system is not critical, the cutoff will not be affected by system size.

Again, the results were published in *Physical Review Letters*. Hans Jakob had managed to become a coauthor of two articles published in the world's most prestigious physics journal before graduating from high school. So if any of my readers should happen to have ideas of their own, don't be shy. Go full speed ahead, and don't let any professional scientist intimidate you.

The model was still very simplified. When there is a rupture in any solid material, not only the nearest neighborhood is affected. In reality, the elastic forces extend to very large distances. Taking this into account, Kan Chen, Sergei Obukhov, and I constructed a much more elaborate model of fracture formation. Starting with a nonfractured solid, a fractal pattern of fault zones emerges, together with a power-law distribution of fracture events. This simulation showed that a fractal fault pattern and the Gutenberg–Richter law could both be derived within a single mathematical model. The results are much more in tune with real earthquakes, where the seismic activity is distributed over a large area and not confined to individual faults. Some earthquakes involve interactions between faults, where the rupture along one fault puts pressure on another fault, which then ruptures during the same earthquake.

Rumbling Around Stromboli

Volcanic activity, like that of earthquakes, is also intermittent, with events of all sizes. A team headed by Paolo Diodati of the University of Perugia, Italy has measured bursts of acoustic emission, that is rumbling sounds, in the area around Stromboli in Italy. They placed piezoelectric sensors coupled to the free ends of steel rods tightly cemented into holes drilled into the rocks. One sensor was placed at a distance from the volcano and another was placed nearer to it. The sensors measured the distribution of the strengths of the burst of activity. Figure 23 shows the distribution for the two signals. Although one signal was weaker than the other, the straight lines on the logarithmic plots have the same slope, with an exponent approximately equal to 2. Diodati claimed that this indicates that volcanic activity is an SOC phenomenon.

It seems that the human brain has not developed a language to deal with complex phenomena. We see patterns where there are none, like the Man in the Moon, and the inkblots shown in Rorschach psychological tests. The human mind cannot directly read the boring straight line in logarithmic plots from observation of geophysical phenomena. First we tend to experience phenomena as periodic even if they are not, for example, at gambling casinos and in earthquakes and volcanos. When there is an obvious deviation from periodicity, like the absence of an event for a long time, we say that the volcano

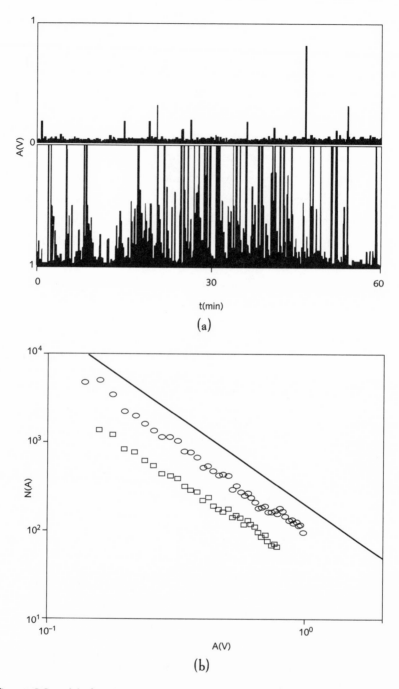

Figure 23. (a) Acoustic emission measured near Stromboli, Italy. The two curves show the strength of the rumbling for a period of one hour as measured at two different distances from the volcano. (b) Size distribution of the bursts of acoustic emission shown in (a). The distribution is a power law with exponent approximately equal to ³/₂, the slope of the straight line (Diodati et al., 1991).

"has become dormant" or the earthquake fault is "no longer active." We try to compensate for our lack of ability to perceive the pattern properly by using words, but we use them poorly. Nothing really happens in an earthquake fault zone in a human lifetime—the phenomenon is a stationary process over millions of years.

The Crust of the Earth Is Critical

In using nitty-gritty computer models, we should not lose track of the greater implications. Because of the robustness under modifications of the models, the criticality does not really depend on our particular choice of model.

The picture that emerges is amazing and simple. The crust of the Earth, working through hundreds of millions of years, has organized itself into a critical state through plate tectonics, earthquake dynamics, and volcanic activity. The crust has been set up in a highly organized pattern of rocks, faults, rivers, lakes, etc., in which the next earthquake can be anything from a simple rumble to a cataclysmic catastrophe. The observations summarized by the Gutenberg–Richter law are the indications that this organizational process has indeed taken place.

So far, we have been viewing earthquakes, volcanic eruptions, river network formation, and avalanches causing turbidite deposition as separate phenomena, but they are all linked together. Earthquakes cause rivers to change their pattern. In Armenia after the 1988 earthquakes near Spitak, a small river had suddenly found a new path through the rocky landscape, and was displaced hundreds of meters from the original riverbed. The shift was not caused by erosion, as is usually the case. Also, it has been proposed that rare external events occurring over a large region, for example, earthquakes or storms, are the dominant source of the turbidite deposits, i.e., the aggregation of material at the continental shelf is not caused by a smooth transport process. The distribution of turbidite deposits simply mirrors the statistics of earthquakes.

In the final analysis, the crust of the Earth can probably be thought of as one single critical system, in which the criticality manifests itself in many different ways. The sandpile theory explains only one level in a hierarchy. The sand must come from somewhere else—maybe another critical system—and it must go somewhere else—perhaps driving yet another critical system. The sandpile describes only one single step in the hierarchical process of forming complex phenomena. Similarly, the crustal plates are fractal structures themselves, indicating that they originate from another critical process, possibly associated with the convective motion of the material in the earth's interior.

Pulsar Glitches and Starquakes

Self-organized criticality is not confined to the Earth, but can be found elsewhere in the universe. A possible example is a pulsar, which is a rotating star consisting of neutrons. Sometimes the star's rotational velocity changes suddenly. These changes in velocity are called *pulsar glitches*. Some of the glitches are small, corresponding to a small change of velocity; some are large, with a large change of velocity.

Ricardo Garcia-Pelayo of the University of Texas in Austin and P. Morley of the Ilya Prigogine Center in Austin made an interesting observation. Using data collected over twenty five years, they created a histogram of the number of pulsar glitches of each size, and discovered that pulsar glitches also follow the Gutenberg–Richter law (Figure 24). They suggested that the pulsar glitches are due to "starquakes" operating in the following way. The surface of the pulsar is under enormous pressure from gravity. Sometimes the crust yields to this pressure, and part of it collapses. Morley and Garcia-Pelayo call this collapse a "starquake." A starquake causes the velocity of the rotations of the pulsar to increase, just as the rotation velocity of an ice skater increases when she draws in her arms. A small starquake causes a small increase in frequency, and a large starquake causes a large increase. The size of pulsar glitches thus reflects the underlying size of the starquake. Morley has constructed a theory of the collapse of the pulsar. Of course, we know much less about pulsars than about our own planet, so the modeling is quite speculative.

Black Holes and Solar Flares

Black holes are massive objects from which nothing can escape, not even light, so we know about their existence only from observation of the gravitational

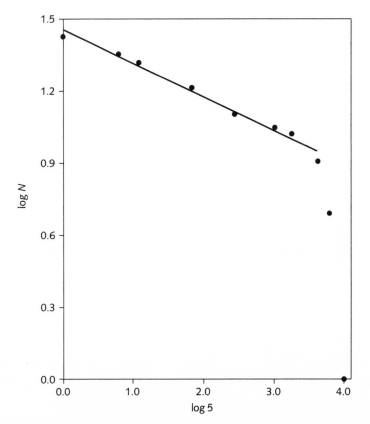

Figure 24.   Gutenberg–Richter law for pulsar glitches (Garcia-Pelaya and Morley, 1993).

force of the black hole working on other cosmic objects. A black hole attracts massive particles from its environment, which are sucked into the interior of the hole, never to be heard from again.

   Recently, Mineshigi, Takeuchi, and Nishimori, in Japan, suggested that this process works very much like a sandpile. The material is temporarily arranged in disks surrounding the black hole. Gas particles are randomly injected into these accretion disks from the environment. When the mass density of the disk exceeds some critical value, the accumulated material begins to drift inward as an avalanche, thereby emitting x-rays that can be observed from the earth. We might think of the process as an hourglass in which sand is falling through a hole in the botton, while new sand is supplied from the outside. The fluctuations of the intensity of the x-rays have a $1/f$ spectrum. On the basis of observations of x-rays from the black hole Cygnus X-1, and some simple computer modeling, the authors conclude that the formation of black holes is an SOC phenomenon.

However, we don't have to travel so far out in the universe to find sources of x-rays with power law distribution. One of the finest and most spectacular applications of the idea involves solar flares. In contrast to pulsars and black holes, we can directly observe what is going on without too much guessing. The sun emits solar flares all the time. Most flares are relatively small. Some of them are very large, but much rarer, and cause disruptions of radio communication on earth.

Solar flares are observed to have a large dynamical range in both energy and duration. The solar flares emit x-rays, so the intensities of the solar flares can be measured as the intensity of these x-rays. Figure 25 shows the distribution of x-rays as measured by instruments on one of NASA's spacecrafts, as presented by B. R. Dennis. The diagram shows the frequency of flares versus their intensity, as given by the measured "count rate." Note the straight-line behavior over more than four orders of magnitude. The flattening of the curve for small flares might well be due to difficulties in measuring these small flares in the background of x-rays from other sources. The slope of the straight line, that is, the exponent τ for the corresponding power law distribution, is approximately 1.6.

A couple of years ago I presented these data at a scientific colloquium on self-organized criticality at the Goddard Space Center in Maryland. A member of the audience rose and said, "I am the Dennis who made this plot. Actually, we have now much more data, and you can extend the straight line scaling regime over another two orders of magnitude."

The physics of solar flares is extremely complicated. The flares are associated with magnetic instabilities in the plasma forming the sun. There has been a good deal of theoretical effort to understand the basic mechanisms. The convective motion of the gas pumps energy into the magnetic field at a steady rate. At some point, there is an instability leading to a breakdown of the pattern of magnetic field lines, which can be viewed as a sudden reconnection of the field lines, like a knot on a shoelace that is released by cutting the lace and gluing the ends together again. The reader might find it difficult to understand this picture. Don't despair—I don't understand it either.

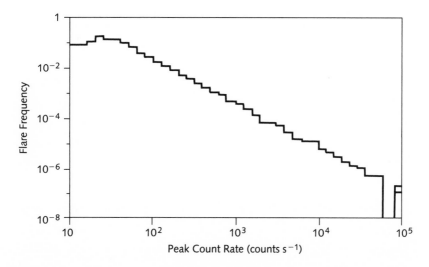

Figure 25. Histogram of x-ray intensity from solar flares, as measured by the NASA satellite ISEE 3/ICE (Dennis, 1985). The diagram shows the relative amount of flares with a given energy, as represented by the "counting rate." The data fit a straight line over four orders of magnitude. The statistics is poor for the few large events.

Lu and Hamilton have constructed a simple theory of the solar flares based on this type of physics. The local magnetic instability can be thought of as the toppling grain of sand that triggers an avalanche of further magnetic instabilities in the solar corona. This avalanche is the solar flare that we observe. Lu and Hamilton constructed an extremely simple model, which has many similarities with the sandpile models and the earthquake models studied by Feder, Olami, and Christensen. The surface of the sun was represented by a grid. On each square on the grid, a field F was defined. In contrast to earthquake models, the field is a vector field, like an arrow with components in three perpendicular directions. Indeed, pictures of the surface of the sun show a grainy texture, much like the sand in the sandpile. The driving of the system was represented by adding small random components to the vectors at a very slow rate. When the "slope," that is the difference between the magnetic field at one site and the average of the field at the six neighboring sites, exceeds some critical value, there is a magnetic breakdown. The breakdown is represented by readjusting the vector fields at the unstable site and the neighboring sites, so that the local configuration becomes stable. However, this rearrangement can cause the slope at nearby sites to exceed the stability limit and cause further breakdown.

Amazingly, this simple theory can explain the satellite observations almost exactly. Lu and Hamilton calculated many different physical properties, including the energy distribution of the flares and the durations of flares with a given energy. All of their results agreed with the satellite data. For instance, the exponent for the energy distribution was found to be $\tau = 1.52$, which compares well with the measured exponent of 1.6.

Lu and Hamilton could draw one simple conclusion: the corona of the sun is working at the self-organized critical state. The theory explains why huge solar flares that disrupt telecommunications occur, on average, every 10 to 20 years. The large events are not periodic, but have statistics similar to large earthquakes and mass extinction events in evolution. These events are at the extreme right end of the tails of the observed distributions. If we have patience enough, we are bound to experience even larger flares with more devastating effects, with a frequency given by extrapolating the critical behavior further.

the "game of life": complexity is criticality

So far we have visited many phenomena on Earth and in the universe. However, one geophysical phenomenon was left out, the most complex of all, namely biological life. In the early days of self-organized criticality, we did not think about biology at all; we had only inert dead matter in mind. However, this situation has radically changed. The story is one in three acts, to be told in the next three chapters, with more to follow. We have constructed some simple mathematical models for evolution of an ecology of interacting species. However, to appreciate the content of the theory that came out at the end, a historical account of the activities seems most suitable.

Our first act is a prologue that deals, not with life, but with the "Game of Life," a toy model of the formation of organized, complex, societies. We showed that the game operates at, or at least very near, a critical state. The second part is very confusing and frustrating, dealing with endless and rather fruitless discussions and collaborations with other scientists on complex phenomena. This work has been associated with the Santa Fe Institute in New Mexico. This is not intended to be a sad story. It only illustrates that the road leading to new scientific insight can be convoluted. Often, initial research is wrong, or at least the signal-to-noise ratio may be very low. The Santa Fe Institute has been sharply criticized, mostly because the institute has been very open to outsiders and has admitted science writers into the process at a preliminary stage, before solid results have been obtained.

After walking around in the dark, eventually there was light at the end of the tunnel. As we shall see in Chapter 8, models relevant to biology do evolve to the critical state. Not only that—some of the models are simple enough that many aspects can be rigorously understood from pen and paper analysis without relying entirely on computer simulations.

The Game of Life is a *cellular automaton*. More than anyone, Stephen Wolfram, then at the Institute for Advanced Studies, Princeton, pointed out that these simple devices could be used as a laboratory for studying complex phenomena. Cellular automata are much simpler than the continuous partial differential equations usually used to describe complex, turbulent phenomena, but presumably their behavior would be similar. Cellular automata are defined on a grid similar to the one on which our sand model is defined. Wolfram mostly studied one-dimensional lattices, but cellular automata can be defined in any dimension. On each point of the grid, there is a number that can be either 0 or 1. At each time step, all the numbers in all the squares are updated simultaneously according to a simple rule. In one dimension, the rule specifies what the content of each cell at the next time step should be, given the state of that particular cell, and its neighbors to the left and to the right, at the present time. The rule could be, for instance, that the cell should assume 1 if two or more of those three cells are 1 otherwise 0.

One can show that in one dimension there are $2^8 = 256$ such rules. Starting from, say, a random configuration of 0s and 1s, some rules lead to a boring state, in which the numbers freeze into a static configuration after some time. Sometimes the rules lead to a "chaotic" state, in which the numbers will go on changing in a noisy way without any pattern, like a television channel where there is no signal. Sometimes, the rule leads to regular geometrical patterns. Wolfram speculated that there was a fourth class that unfortunately was never defined (and therefore not found), in which the automaton would produce new complex patterns forever.

It has now been demonstrated by computer simulations, in particular by Dietrich Stauffer of the University of Cologne, Germany, that none of the one-dimensional automata show truly complex behavior; they can all be classified into the first three classes.

Wolfram never produced any theory of cellular automata. Eventually, he left science completely, and went on to form a computer software company, whose greatest achievement is the program Mathematica for automatic manipulation of mathematic expressions. Wolfram often expressed the view that the automata could be "computationally irreducible," or undecidable, which means that the only way to find the outcome of a specific rule with a specific initial condition is to simulate the automaton on a computer. However, while this view might seem like the end of the story for a mathematician, this does not prevent the physicist from a statistical, probabilistic description of the phenomenon. Many problems that physicists deal with, such as dynamic models of phase transitions, might well be undecidable. The problem of computational irreducibility doesn't keep the physicists awake at night, since there are approximate methods available that give eminently good insight into the problem.

In two dimensions, there is an even richer world than in one dimension. Often, the neighborhood that is considered when updating a site is restricted to eight neighbors—those at the left, right, up, and down positions, and those at the four corners at the upper and lower left and right positions—and to the site itself. There are a total of 2^{512} possible rules specifying how to update a cell, that is a number written as 1 followed by more than 150 zeros. It is obviously impossible to investigate them all, even with a computer.

Many years before Wolfram, the mathematician John Horton Conway of Princeton University had studied one of these zillions of two-dimensional rules called the Game of Life. Presumably he was trying to create a model of the origin of complicated structures in societies of living individuals. Although the Game of Life has never been taken seriously in a biological context, it has nevertheless served to illustrate that complex phenomena can be generated from simple local rules. The Game of Life was described in a number of classic articles by Martin Gardner in *Scientific American* in the beginning of the 1970s. Gardner involved his readers in an exciting hunt for amazingly complicated and fascinating phenomena in this simple game.

The game is played on a two-dimensional grid as follows. On each square, there may or may not live an individual. A live individual is represented by a 1, or a blue square in Plates 6 through 8. The absence of an individual is represented by a 0, or a light gray square. At each time step, the total number of live individuals in the nine-cell neighborhood of a given cell is counted. If that number is greater than 3, an individual at that cell dies, presumably of overcrowding. If the number is 1 or 0, he will die of loneliness. A new individual is born on an empty square only if there are precisely three live neighbors. The red sites are empty sites where a new individual will be born at the next time step. In the figures, individuals who are going to die at the next update are shown as green squares, and empty sites where an individual will be born are shown as red squares. Notice that each red site indeed is surrounded by precisely three blue cells among its eight neighbors.

A myriad of complicated structures can be constructed from these rules. The figures show some stable blue clusters of live individuals. Note that the number of live neighbors in the neighborhood of each live site is either 2 or 3. There are also configurations that propagate through the lattice. The simplest is the *glider*, shown in Plate 6 near the lower right corner. In a small number of time steps, the glider configuration reproduces itself, at a position that is shifted in a diagonal direction of the grid. It keeps moving until it hits something. The gray areas show where there has been recent activity, so the path of the glider is shown as a gray trail behind it. "Blinkers" shift back and forth between two states, one with three individuals on a horizontal line, the other with three individuals on a vertical line. The blinking comes about by the death of two green sites and the simultaneous birth at two red sites. There are more complicated formations involving four blinkers, as shown in Plate 6. There are incredibly ingenious configurations, such as *glider guns*, which produce gliders at a regular rate and send them off in the diagonal direction. There are even structures that bounce gliders back and forth. The number and variety of long-lived structures in the Game of Life is evidence of its emergent complexity. Conway's interest in the game was in its ability to create this fascinating zoo of organisms.

Michael Creutz is a particle physicist at Brookhaven Laboratory, best known for his "lattice gauge theories" in particle physics. In 1988 Mike thought of applying computational methods to relativistic quantum field theory, the current theory of elementary particles. The particles were described by a statistical sampling over a three-dimensional grid, rather than in continuous space, to make the problem computationally tractable. Kan Chen, Michael Creutz, and I became interested in the Game of Life. Our interest was not in "stamp collecting" all the complex structures, but in the general understanding of what makes the Game of Life tick. What is special about the particular rule that Conway had chosen?

If one starts the game from a totally random configuration of live individuals, the system will come to rest after a long time in a configuration in which there are only stable static structures and simple blinkers. All moving objects, such as the gliders, will have died out. It appeared to us that the Game of Life might operate at a critical state. To test this hypothesis, we made a careful computer simulation.

We started from a random configuration and let it come to rest in a static configuration. Such a static configuration, with stable clusters and blinkers only, is shown in Plate 7. We then made a single "mutation" in the system, by adding one more individual, or removing one at a random position. This is analogous to the addition of a single grain to the sandpile model at a random position. The addition of a single individual may cause a live site to die because the number of live individuals in its neighborhood becomes too large. It may also give rise to the birth of a new live site by increasing the number of live neighbors of dead sites from 2 to 3. This creates some activity of births and deaths for a while, where new clusters of living objects are coming and going, and gliders are moving back and forth. Eventually the system again comes to rest at another configuration with static objects, or simple periodic blinkers, only. Then we would make another mutation, and wait for the resulting disturbance to die out. Sometimes the Game of Life comes to rest after a small number of extinction and creation events, sometimes after a large number of events.

We repeated the process again and again. The process that starts when a new individual is added or removed and stops when a static configuration is reached is called an avalanche. The size s of the avalanche is the total number of births and deaths occurring before the avalanche stops. The duration t of the avalanche is the total number of time steps involved. The size s is greater than the duration t because at each time step there are usually many births and deaths taking place simultaneously. Plate 8 shows a snapshot of an avalanche in progress. The gray area indicates sites where at least one individual was born or died during the avalanche.

Because of the magnitudes of the largest avalanches, which involved up to 100 million births and deaths, the computations were very time-consuming. Amazingly, we found ourselves, supposedly serious scientists at a prestigious national laboratory, playing computer games for hundreds of hours on the biggest mainframe computer at the lab.

The distribution was the usual power law, shown in Figure 26. The exponent, measured in the usual way as the slope of the curve, is $\tau = 1.3$. This shows that the Game of Life is critical! Surprisingly, one can make a theory for this value of τ based on a connection of the Game of Life to sophisticated theories of particle physics, as we shall see in Chapter 9. The number of time steps follows another power law, with an exponent that can be calculated from the same theory. This mind-boggling connection was found a few years later by Maya Paczuski, a Department of Energy research fellow working at Brookhaven.

Many other computer simulations have been performed, following our original work. Some of this work was performed on massive parallel computers with capacity well beyond ours. Some researchers, including Preben Alstrøm of the Bohr Institute in Denmark, confirmed our result of criticality. Others claimed that there actually is an upper cutoff of the size of avalanches; Jan Hemmingsen at the Julich Research Institute in Germany found no avalanches exceeding one billion flips, but there are so few of those large avalanches that the statistics are too poor to make firm claims. In any case, the system is extremely close, within 1 part in 100 million, to being critical. However, if these latter scientists are right, it might be an incredible accident that the Game of Life is critical. Dietrich Stauffer investigated systematically millions of two-dimensional cellular automata and was not able to find a single additional critical model. This indicates that the Game of Life does not exhibit robust criticality. If you change the rules, you destroy the criticality.

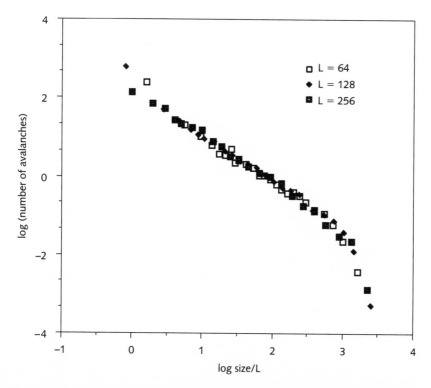

Figure 26.   Distribution of avalanche sizes in the Game of Life. The curves for different lattice L cover each other when plotted vs. s/L rather than L. This demonstrates finite size scaling, indicative of critical behavior. The exponent of the power law is $\tau = 1.30$.

Self-organized critical systems must be precisely critical without any tuning. If the criticality in the Game of Life is not self-organized, then it is accidental. John Conway must have tuned it to be extremely near to criticality. Conway is the watchmaker in the Game of Life! We don't know how much Conway experimented before he arrived at the Game of Life, unique among millions of millions of other rules. He was interested in the endogenous complexity of his creatures. But our calculations show that at the same time that he had succeeded in constructing something that exhibited a vast amount of complexity, he had (inadvertently, I guarantee) tuned the system to be critical! At the time of Conway's work, little was known about the concept of critical phenomena even in equilibrium systems, so Conway cannot have known anything about criticality. Among those many possible rules, he had arrived precisely at the one that is critical. I still wonder what in the world made him hit upon this absurdly unlikely model, in view of the fact that the world's largest computers have not yet been able to come up with another complex one.

Only the critical state allows the system to "experiment" with many different objects before a stable complex one is generated. Supercritical, chaotic rules will wash out any complex phenomenon that might arise. Subcritical rules will freeze into boring simple structures.

The message is strikingly clear. The phenomena, like the formation of the "living" structures in the Game of Life, that we intuitively identify as complex originate from a global critical dynamics. Complexity, like that of human beings, which can be observed locally in the system is the local manifestation of a globally critical process. None of the noncritical rules produce complexity. *Complexity is a consequence of criticality.*

is life a self-organized critical phenomenon?

The step from describing inert matter to describing biological life seems enormous, but maybe it isn't. Perhaps the same principles that govern the organization of complexity in geophysics also govern the evolution of life on earth. Then nature would not suddenly have to invent a new organizational principle to allow live matter to emerge. It might well be that an observer who was around when life originated would see nothing noteworthy—only a continuous transition (which could be an "avalanche") from simple chemical reactions to more and more complicated interactions with no sharp transition point indicating the exact moment when life began. Life cannot have started with a chemical substance as complicated as DNA, composed of four different, complicated molecules called nucleotides, connected into a string, and wound up in a double helix. DNA must itself represent a very advanced state of evolution, formed by massively contingent events, in a process usually referred to as prebiotic evolution. Perhaps the processes in that early period were based on the same principles as biology is today, so the splitting into biotic and prebiotic stages represents just another arbitrary division in a hierarchical chain of processes.

Maybe a thread can be woven all the way from astrophysics and geophysics to biology through a continuous, self-organized critical process. At this time all the intermediate stages of evolution progressing from chemistry to life are distant history, so we see geophysics and biology as two separate sciences.

Biology involves interactions among millions of species, each with numerous individuals. One can speculate that the dynamics could be similar to that of sandpiles with millions of interacting grains of sand. However, the realization of this idea in terms of a proper mathematical description is a long and tedious process. Much of my thinking along these lines took place at the Santa Fe Institute, mostly through interactions with Stuart Kauffman, who resides there. For three years Stuart and I were walking around in circles without being able to make a suitable model of evolution, but eventually this work paid off in a rather surprising turn of events.

The Santa Fe Institute

The Santa Fe Institute in New Mexico is a lively center for exchange and debate on complex systems. In the words of the economist Brian Arthur of Stanford University, now the Citibank Professor at the institute, "It is the only place where a biologist can come and hear an economist explain how a jet engine works." The institute brings together many of the most imaginative thinkers from vastly different fields in an open environment. The meetings at Santa Fe are continuous brain storms.

The institute is the brainchild of George A. Cowan, former head of research at Los Alamos National Laboratory near Santa Fe. It soon received the backing of top scientists in a number of fields, including Philip W. Anderson, Nobel Prize winner for his work on condensed matter physics, Murray Gell-Man, Nobel Prize winner for the discovery of quarks, which are among the most fundamental of all particles, and Kenneth Arrow, economist and Nobel Prize winner for the general equilibrium theory of economics.

The reductionist approach has always been the royal road to the Nobel Prize. Ironically, the philosophy of the institute is quite orthogonal to the re-

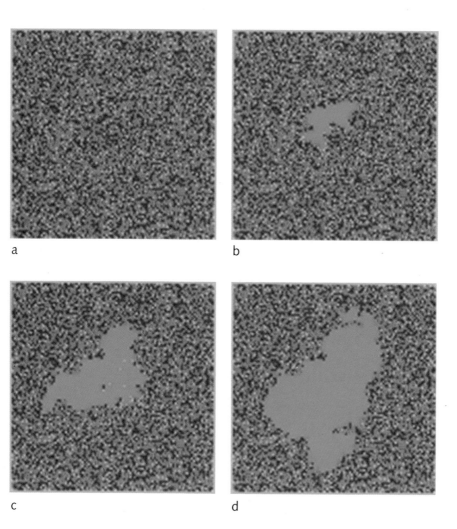

Plate 1. Snapshots of propagating avalanche in the sandpile model. The colors gray, green, blue, and red indicate heights of 0, 1, 2, and 3, respectively. The light blue show the columns that have toppled at least once. As the avalanche grows, the light blue area increases. (Courtesy of Michael Creutz.)

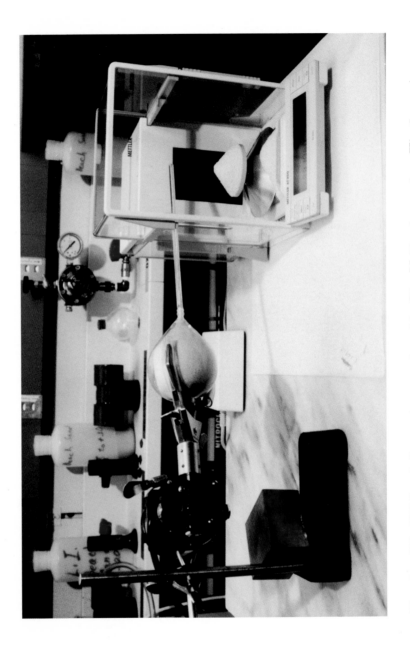

Plate 2.   The IBM sandpile experiment, performed by Glen Held and co-workers. The fluctuating mass of the sandpile on the scale is analyzed by a small PC.

a

b

Plate 3.   Sandpile experiments by the University of Michigan group led by Michael Bretz and Franco Nori. (a) Tilted sandpile. (b) Conical pile. The sandpile shown is the digital image from the video recorder.

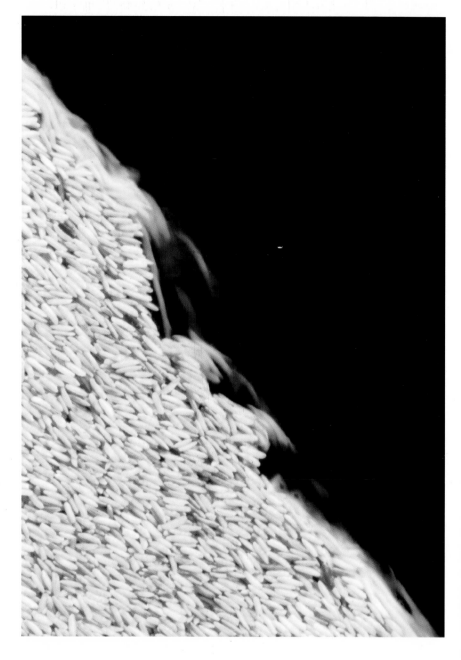

Plate 4.   Rice pile profile in the self-organized critical state. (Frette et al., 1995.)

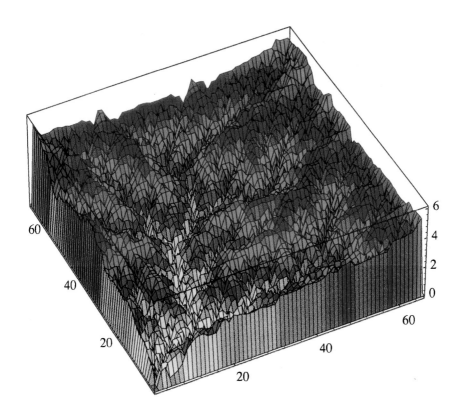

Plate 5. The self-organized fractal landscape corresponding to the river network in Figure 20. The colors from yellow to green, blue, and cyan reflect increasing elevation. (Rigon et al., 1994).

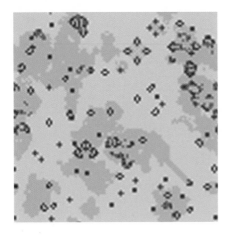

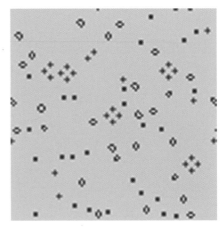

Plate 6. Running configuration of the Game of Life. Blue sites are stable live sites. The red sites are empty sites that will become live at the next update. The green sites are live sites that are moribund. The darker gray are sites with recent activity. Note the glider near the lower right corner, leaving a gray track behind it.

Plate 7. Static configuration in the Game of Life, with stable clusters and blinkers. Note also the formation of clusters of blinkers. (Courtesy of Michael Creutz.)

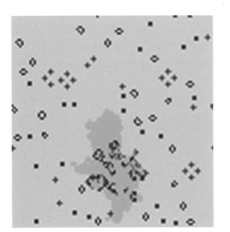

Plate 8. Avalanche propagating in the Game of Life, starting from the static configuration shown in Plate 7. The avalanche was initiated by adding a single live site. The gray area has been covered by the avalanche, so the configuration within that area is different from that of Plate 7.

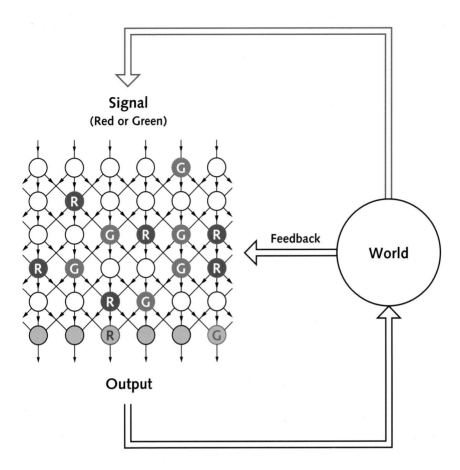

Plate 9. Block diagram of brain interacting with the outer world. The world shows a red or green signal to the brain. The signal is fed into the brain at arbitrary neurons. The bottom row represents the action resulting from the processes going on in the brain. This is transmitted to the environment, which provides a feedback. If the response is correct, the environment provides food; if the response is not correct, the environment does not provide food. (Stassinopoulos and Bak, 1995).

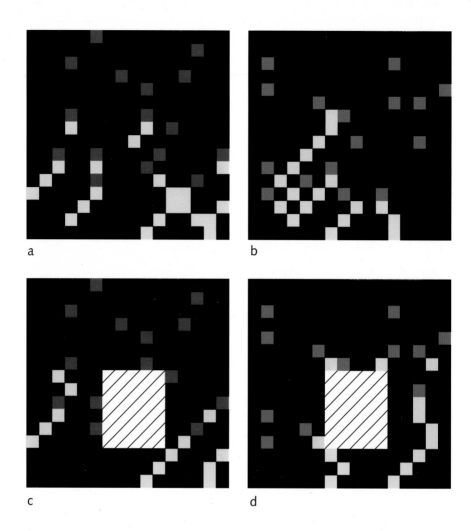

Plate 10. Successful firing patterns in the fast switching phase. The two sets of input neurons are colored red and green, respectively. For (a) the red input in the cells #10 and #25 of the bottom row must be triggered; for (b) the green signal in the output cells #7 and #12 must be triggered. The yellow squares show which outputs are firing for the two inputs . Successful patterns after the removal of a block of thirty neurons (c) and (d). Note the difference from the original response. (Stassinopoulos and Bak, 1995).

ductionist science which propelled those gentlemen to stardom. Complexity deals with common phenomena in different sciences, so the study of complexity benefits from an interdisciplinary approach. However, because of the sociology of science, it takes someone at the top to change the course of science. Most scientists in the rank and file do not venture into new areas that have not been approved from above. There is good reason for this since young scientists are dead in the water if they step out of traditional disciplines.

Traditionally, cross-disciplinary research has not been very successful. The fundamental entities dealt with in the various sciences are completely different: atoms, quarks, and strings in physics; DNA, RNA, and proteins in biology; and buyers and sellers in economics. Attempts to find common ground have often been contrived and artificial. At universities, the different sciences are historically confined to specialized departments with little interaction. This has left vast areas of science unexplored. However, a new view is emerging that there could be common principles governing all of those sciences, not directly reflected in the microscopic mechanisms at work in the different areas. Maybe similarities arise due to the way the various building blocks interact, rather than to the way they are composed.

Since the Santa Fe Institute does not have a permanent staff of scientists, it can change its emphasis quickly when new ideas come up. A number of external professors are associated with the institute; I am fortunate enough to be one. In contrast, traditional university and government laboratory environments have a tendency to freeze into permanent patterns as their scientists become older. Typically, a couple of long-term visitors, some short-term visitors, and a few young postdoctoral fellows work at the institute. In addition, scientists from various fields come together at seminars and conferences.

These meetings force us to place science in a greater perspective. In our everyday research, we tend to view our own field as the center of the world. This view is strengthened by our peer groups, which are, because of the compartmentalization of science, working along the same line. No mechanism for changing directions exists, so more and more efforts go into more and more esoteric aspects of well-studied areas that once paid off, such as high-temperature superconductivity, surface structures, and electronic band structures, without any restoring force. Nobody has an incentive to step back and ask himself, "Why am I doing this?" In fact, many scientists are put off if you ask this question.

This state of affairs struck me at one of the meetings arranged by the institute. Brian Goodwin, a biologist from England with his own view of biologic evolution expressed in his book *How the Leopard Got Its Spots,* had invited twenty scientist to a meeting on "Thinking about Biology." Who did he invite? A couple of biologists, two engineers, some computer scientists and mathematicians, a medical doctor, and some physicists, including me, and some individuals who could not be placed in any category. Goodwin is not in the mainstream of biology—otherwise he wouldn't be at the institute, but would probably be working hard on a molecular biology problem at his home institution.

"What the heck is this all about?" I asked upon arriving. "You are arranging a meeting entitled 'Thinking about Biology,' so why don't you invite someone who is actually thinking about biology?" "This is it!" Brian exclaimed. "There is essentially nobody else thinking about the fundamental nature of biology."

How can that be? In physics at that time (and probably even now) there were an estimated 15,000 scientists working on high-temperature superconductivity, a subject of some general interest and possible technological importance, but nothing that would justify this level of activity. At the same time, only a scattered handful of oddballs were working on understanding life itself, perhaps the most interesting of all problems.

My first visit to the institute was a couple of years before this biology meeting, in the fall of 1988. I was called by one of my physics colleagues, Richard Palmer of Duke University. "We are a couple of people interested in your ideas on sandpiles," he said. "Brian Arthur is running a program here on economics, and he would like to invite you to come." Economics? What did I know about economics?

The institute was about to change my views of science, and came to affect my research profoundly. I fell in love with the place immediately. Discussions would take place outdoors in a little courtyard in the center of the institute, or at one of the many nearby New Mexican restaurants. Numerous informal, but loud, discussions, on life, the universe, and everything else took place at the Canyon Cafe, our "faculty club."

The program was not really about economics, but about common problems in many sciences including biology, geophysics, and economy. Stuart Kauffman, originally a medical doctor but now working on myriad fundamental issues in biology is the heart of the institute. I soon learned that Stu is a unique scientist: fun, playful, and imaginative. Stu is one of the few biologists who are willing and able to view things in an abstract way, to view reality as just one example of a general process.

I gave a short informal presentation of our sandpile model, and our simulations of the Game of Life. Our article was about to appear in *Nature*. In particular, I jokingly put forward the speculation that real life operates at the critical point between order and chaos.

Sandpiles and Punctuated Equilibria

In 1989, I returned to the institute for another month. "I have really been looking forward to meeting you again," Stu exclaimed, putting his hand on my shoulder. "You won't believe how far we have taken your ideas of sandpiles."

And then he told me about Stephen Jay Gould and Niles Eldridge's ideas of "punctuated equilibria" in evolution. Punctuated equilibrium is the idea that evolution occurs in spurts instead of following the slow, but steady path that Darwin suggested. Long periods of stasis with little activity in terms of extinctions or emergence of new species are interrupted by intermittent bursts of activity. The most spectacular events are the Cambrian explosion 500 million years ago, with a proliferation of new species, families, and phyla, and the extinction of the dinosaurs about sixty million years ago. The evolution of single species follows the same pattern. For long periods of time, the physical properties, like the size of a horse or the length of the trunk of an elephant, do not change much; these quiet periods are interrupted by much shorter periods, or punctuations, where their attributes change dramatically. Darwin argued and believed strongly that evolution proceeds at a constant rate.

Indeed, sandpiles exhibit their own punctuated equilibria. For long periods of time there is little or no activity. This quiescent state is interrupted by rapid bursts, namely the sand slides, roaming through the sandpile, changing

everything along their way. The similarity between the avalanches in the sandpile and the punctuations in evolution was astounding. Punctuations, or
avalanches are the hallmark of self-organized criticality. Not long after my first
visit, Stu had plotted Sepkoski's data for extinction events in the evolutionary
history of life on earth the same way we had done it for the sandpiles, and found
that the data were consistent with a power law, with the large extinction events
occurring at the tail of the distribution (Figure 5). Could it be that biological
evolution operates at the self-organized critical state? The idea had enormous
implications for our views of life on earth. Life would be a global, collective, cooperative phenomenon, where the complex structures of individual creatures
would be manifestations of the dynamics of this critical state, just like the organisms in Conway's Game of Life. But how could one express the idea in a theoretical framework, in view of the inherent difficulties that were encountered
when modeling a system as straightforward as a sandpile?

Interacting Dancing Fitness Landscapes

Before going further let us take a look at the important concept of "fitness
landscapes," described by Sewall Wright in a very remarkable article, "The
Shifting Balance Theory," from 1952 (reviewed in Wright's 1982 article). The
physical properties of biological individuals, and thus their ability to survive
and reproduce, depend on "traits" of that individual. This ability to survive
and reproduce is referred to as "fitness." A trait could be the size of the individual, the color or the thickness of the skin, the ability of the cell to synthe-
size certain chemicals, and so on. The traits express the underlying genetic
code. If there is a change of the genetic code, that is a change in the genotype,
there may or may not be a change of one or more of these traits, that is a change
in the physical appearance or phenotype, and therefore a change in fitness.

Wright suggested that fitness, when viewed as a function of the many-
dimensional trait-space with each dimension representing a trait, forms a
rough landscape, as illustrated in Figure 27. Since the traits reflect the underlying genes, one might think of the fitness as being a function of the genetic
code, represented by the little black and white squares. Some genetic combinations correspond to particularly fit individuals and are shown as peaks in

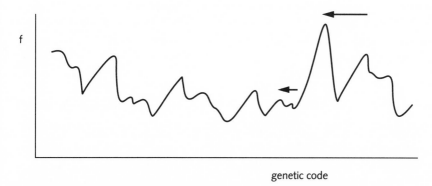

genetic code

Figure 27.   Sewall Wright's fitness landscape. Note that the species located at low-lying peaks have smaller barriers (valleys) to cross to improve their fitness than the species located at the high peaks.

the diagram. Some other combinations give rise to individuals with little viability, and are represented by valleys. As the genetic code is varied over all possible combinations, the fitness curve traces out a landscape. There are numerous tops and valleys corresponding to the many very different possibilities of having fit (and unfit) genes. A mutation corresponds to taking a step in some direction in the fitness space. Sometimes that will be a step down, to a state with lower fitness, and sometimes that would be a step up, to a state with higher fitness.

A species can be thought of as a group of individuals localized around a point in the fitness landscape. In the following discussion I will take the liberty of representing an entire species population in terms of a single point, which I will refer to as the "fitness of the species." Each individual member of a species undergoes random mutations. The fitter variants, by definition, will have larger survival rates, and will proliferate and take over the whole population. Downhill steps will be rejected, uphill steps accepted. Hence by random mutation and selection of the fitter variants, the whole species will climb uphill. At this level there is not much difference between Darwin's selection of fitter variants among random mutation and Lamarque's picture of evolution as being directed toward higher fitness—it is only a matter of time scales. Both lead to hill climbing. Darwin's theory provides a mechanism for Lamarque's directed evolution. In other words, even if Lamarque was wrong and Darwin was right, this may not have any fundamental consequences for the general structure of macroevolution.

Many early theories of evolution, including Fisher's celebrated work, *The Genetical Theory of Natural Selection* of the 1930s, can be understood simply as a detailed description of this uphill climbing process in a situation where the mountains have a constant slope, and are infinitely high. The fitnesses increase forever, implicitly representing the view that evolution is progress. Fisher's math didn't even touch the complexity and diversity of evolution—everything was neat and predictable.

Unfortunately, there is a pervasive view among biologists that evolution is now understood, based on these early theories, so that there is no need for further theoretical work. This view is explicitly stated even in Dawkins' book *The Blind Watchmaker.* Nothing prevents further progress more than the belief that everything is already understood, a belief that has repeatedly been expressed in science for hundreds of years. In all fairness, all that Dawkins is saying is that Darwin's mechanism is sufficient to understand everything about evolution, but how do we know in the absence of a theory that relates his mechanism at the level of individuals to the macroevolutionary level of the global ecology of interacting species?

In Sewall Wright's picture, however, the uphill climb must necessarily stop when the fitness reaches a peak. When you sit on top of a mountain, no matter which direction you go, you will go downhill. If we take a snapshot of biology, we can think of the various species as sitting near peaks in their landscapes. To get from one peak to a better one, the species would have to undergo several simultaneous, orchestrated mutations. For instance, a grounded species would have to spontaneously develop wings to be able to fly. This is prohibitively unlikely. Therefore, in Wright's picture evolution would come to a happy end when all species reach a local maximum. There may be better maxima somewhere else, but there is no way to get there. Evolution will get to a "frozen" state with no further dynamics.

What is missing in Sewall Wright's fitness landscape? Stuart Kauffman suggested that the important omission was the interaction between species. Species affect each other's fitnesses. When a carnivorous animal develops sharper teeth, that reduces the fitness of its prey; vice versa, if the prey develops thicker skin, or if the animal becomes quicker, or if it becomes extinct, that affects the livelihood of its prospective predators. In Stuart's favorite example, if a frog develops a sticky tongue in order to catch a fly, the fly can react by developing slippery feet. Diagrammatically, the interactive ecology can be illustrated as in Figure 28. The squares represent species. An arrow from one species to another indicates that the latter species depends on the physical properties of the first. Sometimes, the arrows point only in one direction. For instance, our body contains numerous viruses and bacteria that benefit from us, but don't affect us. Often the arrows point in both directions when the two species have symbiotic relations to each other, or when a parasite benefits from, but harms, its host. Biology might be thought of as the dynamics of a collection of interactive species living in a global ecology.

The fitness landscapes of the various species are "deformable rubber landscapes" that interact with one another. The landscapes may change.

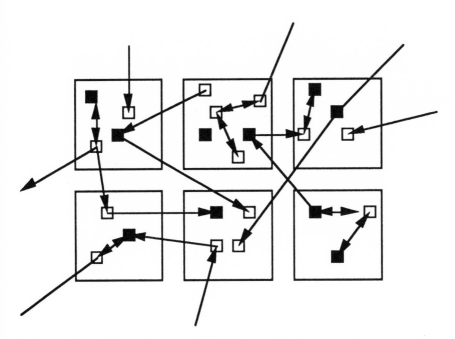

Figure 28.   Diagram of interacting species. The squares represent species in an ecology. An arrow pointing from one species to another indicates that the latter is affected by the former. Sometimes the arrows go in one direction, sometimes in both.

When a species mutates and changes its own properties, it changes the shapes of the landscapes seen by other species. A species living happily on top of one of the hills of its own fitness landscape may suddenly find itself way down the slope of the mountain. But then the species can respond by climbing to a new top, by random mutation and selection of the fitter variant. Using Stuart's metaphoric example: A frog may improve its ability to catch flies by developing a sticky tongue; the fly can respond by developing slippery feet. The fly has to evolve just to stay where it was before. It never actually improves its fitness; it must evolve in order to simply survive as a species.

This is called the Red Queen effect, after a character in Lewis Carroll's *Through the Looking-Glass*. "'Well, in our country,' said Alice, still panting a little, 'you'd generally get to somewhere else—if you ran very fast for a long time as we've been doing.' 'A slow sort of country!' said the Queen. 'Now, *here*, you see, it takes all the running you can do, to keep in the same place.'"

We are living in "the fast place" where you have to run in order to go nowhere, not the slow place with a static landscape. In the absence of interactions between species, evolution would come to an abrupt halt, or never get started in the first place. Of course, the fitness landscapes could change because of external effects, such as a change of climate that would change the landscapes of all species.

The solution is to consider coevolution of interacting species, rather than evolution of individual species in isolation that comes to a grinding halt. Co-evolution of many species can be described conceptually in terms of fitness landscapes that affect one another. Stuart Kauffman calls them "interacting dancing fitness landscapes." This picture is a grossly simplified representation of the highly complicated population dynamics of real species coevolving in the real world, but nevertheless it represents a monumental computational challenge to find the ramifications of this view. It could provide a valuable metaphor. The competition between two species is quite well understood in terms of predator-prey models, but what are the consequences for a global ecology with millions of interacting species?

Stu and Sonke Jonsen, a postdoctoral fellow from Norway, were implementing fitness landscapes in terms of interacting models, called "NKC models." They represented each species by a string of N o's and 1's,

(100000 . . . 111100), representing the states of N genes or traits. In the simplest version of the model, they would associate a random number to each of the 2^N configurations, representing the fitness of that configuration. A little black square might represent a 1, a white square a 0. The randomness represents our lack of knowledge of the couplings. This version is the same as the "random energy model" introduced by Bernard Derrida of the University of Paris, in a different context. So far, the model represents a single landscape. If one tries to flip a single bit, from 1 to 0 or from 0 to 1, one finds either a lower fitness or a higher fitness. Selecting the higher value represents a single step uphill in the fitness landscape.

Thus, a very complicated process, namely the mutation of a single individual and the subsequent selection of that fitter state for the whole species population, was boiled down to a change of a single number. A single flip corresponds to a "mutation" of the entire population of a given species, or, equivalently, extinction of one species followed by the replacement of another with different properties! Here and in the following discussion this process is referred to as a "mutation of the species."

Many evolutionary biologists, such as John Maynard Smith, the author of the bible on traditional evolutionary thinking, *The Theory of Evolution,* insist on locating the mechanisms of evolution in the individual, and find concepts like species mutation revolting. Of course, the basic mechanisms are operating at the individual level; we are simply using a more coarse-grained description to handle the entire macroevolution. Each step involves many generations. Stephen Jay Gould uses the same terminology in some of his books, precisely to be able to discuss evolution on a larger scale than is usually done by geneticists. Not even the gradualists would question that differential selection of the fitter variant leads to the drift of entire species. It is precisely this drift of species that is eventually described by Fisher's theory. The coarse graining does not in itself produce "punctuated equilibria" since we envision this single step to take place in a smooth, gradual way, just as a single falling grain, containing many individual atoms, does not constitute a punctuation or avalanche in the sandpile. In the final analysis, if using a fine-enough time scale, everything, even an earthquake, is continuous. Punctuated equilibria refers to the fact that there is a vast difference in time scales for the periods of stasis, and the intermittent punctuations. The periods of stasis may be 100 million years, while the duration of the punctuations may be much less than a million years.

Eventually, when the process of selecting the fitter variant is continued for some time, the species will eventually reach a local peak from which it cannot improve further from single mutations. Of course, by making many coordinated mutations the species can transform into something even more fit, but this is very unlikely.

Each species is coupled to a number C of other species, or, more precisely, to one particular trait (which could be decided by one gene) in each of C other species, where C is a small integer number. This situation is described in Figure 27, where the small black and white squares could represent genes that are 1 and 0, respectively. The two genes that are coupled could represent, for instance, the slippery foot of the fly, and the stickiness of the surface of the tongue of the frog. If that particular gene in one of the species flips, the viability of the other interacting species is affected. The fitness of the frog depends not only on its own genetic code, but also on the genetic code of the fly. In the model, this coupling is represented by assigning a new random number to a species if the gene to which it is coupled mutates. The interacting species could either be neighbors on a two-dimensional grid, or they could be chosen randomly among the $N - 1$ other species.

A mathematical biologist should in principle be able to study this type of system by using the much more cumbersome methods of coupled differential equations for population dynamics, called Lotka-Volterra equations, or replicator equations. In those equations, the increase or decrease of the population of a species is expressed in terms of the populations of other species. But the computational costs are so tremendous that it limits the systems that can be studied to include very few interacting species, say two or three. Indeed, the dynamics of coevolution of a small number of species have been previously studied, for instance in the context of predator-prey, or parasite-host coevolution. This is insufficient for our purposes, where the conjecture is that the complexity comes from the limit of many interacting species.

The limit by which the number of species is very large, in practice infinity had never before been investigated. The spirit is the same as for our sandpile or slider block models for earthquakes. Instead of following the details of the dynamics, a coarsened representation in terms of integer numbers is chosen. A species is either there or not. We do not keep track of the population of the species, just as we did not keep track of the rotation angle in our pendulum models.

Because of their simpler, though still enormously complicated structure, Kauffman and Jonsen were able to study the situation in which there was a large number of species, each interacting with C other species. They started from an arbitrary configuration in which each of, say, 100 species were assigned a random sequence of numbers 1 and 0. At each time step, they made a random mutation for each species. If this would improve the fitness of the species, the mutation was accepted, that is a single 1 was replaced by a 0, or vice versa. If the fitness was lowered, the mutation was rejected, and the original configuration was kept.

If the value of C is low, the collective dynamics of the ecology would run only for a short time. The first mutation might knock another species out of a fitness maximum. That species will mutate to improve its fitness. This might affect other species. Eventually, the domino process will stop at a "frozen" configuration where all the species are at the top of a fitness peak, with no possibility of going to fitter states through single mutations. All attempts to create fitter species by flipping a single gene would be rejected at that point. This is similar to the situation with no coupling between species. In theoretical biology such a state is called an "evolutionary stable state" (ESS), and has been studied in great detail by mathematical biologists, in particular by John Maynard Smith. Economists call such states, in which no one can improve their situation by choosing a different strategy "Nash equilibria." There is a rather complete mathematical theory of those equilibria derived within the mathematical discipline known as game theory. However, game theory does not deal with the important dynamical problem of how to get to that state, and where you go once the state ceases to be stable.

If, on the other hand, each species interacts with many other species, that is, C is large, the system enters into a "chaotic" mode in which species are unable to reach any peak in their fitness landscape, before the environment, represented by the state of other species, has changed the landscape. This can be thought of as a collective "Red Queen" state, in which nobody is able to get anywhere. Evolution of the single species to adjust to the ever-changing environment is a futile effort.

Both these extremes are poor for the collective well-being of the system. In one case, species would freeze into a low-lying peak in the fitness landscape with nowhere else to go. "Everybody is trapped in the foothills," Stu explains. In the second case, evolution is useless because of the rapidly varying environment. As soon as you have adjusted to a given landscape, the landscape has changed. There is no real evolution in either of those two cases. This leaves but one choice: the ecology has to be situated precisely at the critical state separating those extremes, that is, at the phase transition between those extremes. Here, the species could benefit from a changing environment, allowing them to evolve to better and better fitness by using the slowly changing environment as stepping stones, without having that progress eliminated by a too rapidly changing environment. "The critical state is a good place to be!" in Stu's words. "There we are, because that's where, on average, we all do best."

This shows a kind of free-market fundamentalist view of evolution. If left to itself the system will do what's best for all of us. Unfortunately, evolution (and the free market) is more heartless than this.

Stu and I worked on various modifications of the model, including models of random glasses borrowed from solid state physics. In a glass, the atoms can sit in many different random arrangements that are stable, just like the species in Kauffman's NKC models. We studied the models in a way that was analogous to the method by which Kan Chen, Michael Creutz, and I had studied the Game of Life. First we would wait for the system to relax to a frozen state. Then we would make one arbitrary additional mutation, and let the system relax again to a new stationary state. Each mutation would generate an avalanche. We were never able to have the system organize itself to the critical point. The result was always the same. The model would converge either to the frozen phase or to the chaotic phase, and only if the parameter C was tuned very carefully would we get the interesting complex, critical behavior. There was no self-organized criticality. Models that are made critical by tuning a parameter, although plentiful, are of little interest in our context.

Despite Stu's early enthusiastic claims, for instance in his book *The Origins of Order,* that his evolution models converge to the critical point, that they exhibit self-organized criticality, they simply don't. Nevertheless, his effort was heroic and insightful. This was the first crude attempt to model a complete biology.

I was in a quite frustrated state. On the one hand we had a picture of self-organized criticality that empirically seemed to fit observations of punctuated equilibria and other phenomena. On the other hand, we were totally unable to implement that idea in a suitable mathematical framework, despite frantically working on the problem. In a collaboration with Henrik Flyvbjerg and Benny Lautrup, theoretical physicists at the Niels Bohr Institute in Denmark, we were even able to prove by rigorous mathematics that the models could never self-organize to the critical point.

However, apart from the question as to what type of dynamics may lead to a critical state, the idea of a poised state operating between a frozen and a disordered, chaotic state makes an appealing picture for evolution. A frozen state cannot evolve. A chaotic state cannot remember the past. This leaves the critical state as the only alternative.

Unfortunately, contrary to Stuart's general worldview and personality, life is not all happiness. In all of our work so far, we had selected a random species for mutation in order to start avalanches. It turned out that all we had to do was to choose the least fit species, which would have the smallest valley in the landscape to jump in order to improve its fitness. After three years of hard work and little progress, persistence finally paid off.

chapter 8

mass extinctions and punctuated equilibria in a simple model of evolution

Darwin's theory is a concise formulation of some general observations for the evolution of life on earth. In contrast to the laws of physics, which are expressed as mathematical equations relating to physical observable quantities, there are no Darwin's equations describing biological evolution in the language of rigorous mathematics, as my colleague and friend Henrik Flyvbjerg once eloquently pointed out. Therefore, it is a highly important matter to determine if Darwin's theory gives an essentially complete description of life on earth, or if some other principles have to be included. Darwin's theory concerns evolution at the smallest scale, microevolution. We do not know the consequences of his theory for evolution on the largest scale, macroevolution, so it is difficult to confront, and possibly falsify, the theory by observations on the fossil record.

It was at the time of Darwin that Charles Lyell formulated the philosophy of uniformitarianism, or gradualism. It was Lyell's view that everything should be explainable in terms of the processes that we observe around us.

working at the same rate at all times. For instance, geological landscape formations are supposed to be formed by smooth processes, and the full scale of events, even those of the greatest extent and effect, must be explained as smooth extrapolations from processes now operating, at their current observable rates and intensities. In other words, the small scale behavior may be extended and smoothly accumulated to produce all scales of events. No new principles need be established for the great and the lengthy processes; all causality resides in the smallness of the observable present, and all magnitudes may be explained by extrapolation.

Darwin accepted Lyell's uniformitarian vision in all its uncompromising intensity. Darwin believed that his mechanism, random mutation followed by selection and proliferation of the fitter variants, would necessarily lead to a smooth gradual evolution. Darwin went so far as to deny the existence of mass extinctions. Since biology is driven by slow and small mutations operating at all times and all places, how can the outcome be anything but smooth? Uniformitarianism underlies many views and opinions in Darwin's *The Origin of Species*, including his hostility to mass extinction. Darwin saw evolution as a slow, gradual process. Darwin claims, "We see nothing of these slow changes in progress until the hand of time has marked the long lapse of ages." This is gradualism in a nutshell.

This view is often shared, without further ado, by many evolutionary biologists. Niles Eldridge, the copromoter of the phenomenon of punctuated equilibria, belongs to that group and concludes that Darwin's theory is incomplete because, Eldridge believes, it cannot explain the catastrophic extinctions. Raup and Sepkoski hold similar views. The external cause could be a change in weather pattern, a volcanic eruption, or an extraterrestrial object hitting the earth. Recently, it has been suggested that cosmic neutrinos from collapse of nearby supernovas, hitting the earth at regular intervals, are responsible. It seems to be a widespread assumption that some cataclysmic impact must be responsible for mass extinction, so the debate has been about which external force was responsible.

To a large degree, Lyell's uniformitarian view is a healthy one. Indeed the microscopic mechanisms are solely responsible for the behavior at all scales. Nothing new has to be introduced at any scale.

However, the uniformitarian theory fails to realize that a simple extrapolation does not necessarily take us from the smallest to the largest scale. A physicist might represent Lyell's philosophy simply as a statement that we live in a linear world. The assumption that a large effect must come from a large impact also represents a linear way of thinking. However, we may be dealing with highly nonlinear systems in which there is no simple way (or no way at all) to predict emergent behavior. We have already seen in different contexts that microscopic mechanisms working everywhere in a uniform way lead to intermittent, and sometimes catastrophic, behavior. In self-organized critical systems most of the changes often concentrate within the largest events, so self-organized criticality can actually be thought of as the theoretical underpinning for catastrophism, the opposite philosophy to gradualism.

Thus, the science of genetics, which might be thought of as the atomic theory of evolution, does not provide an answer to the question of the consequences of Darwin's theory, precisely because we cannot extrapolate directly from the microscopic scale to the macroscopic scale. G. L. Simpson, in his famous book *Tempo and Mode in Evolution* states this observation very explicitly in his introduction:

> [Geneticists] may reveal what happens to a hundred rats in the course of ten years under fixed and simple conditions, but not what happened to a billion rats in the course of ten million years under the fluctuating conditions of earth history. Obviously, the latter problem is much more important.

Stephen Jay Gould uses this argument to justify that only a historical, narrative approach to studying evolution is possible, underlining the importance of his own science, paleontology, which deals with the study of the fossil record. Indeed, such studies are indispensable for providing insight into the mechanisms for evolution on a grander scale.

Our approach is to explore, by suitable mathematical modeling, the consequences of Darwin's theory. Perhaps then we can judge if some other principles are needed. If the theory of self-organized criticality is applicable, then the dynamics of avalanches represent the link between Darwin's view of continuous evolution and the punctuations representing sudden quantitative and qualitative changes. Sandpiles are driven by small changes but they nevertheless exhibit large catastrophic events.

The mathematical models that Stuart Kauffman and I had studied were absurdly simplified models of evolution, and failed to capture the essential behavior. There was no self-organized critical state and no punctuated equilibrium. It turns out that the successful strategy was to make an even simpler model, rather than one that is more complicated. Insight seldom arises from complicated messy modeling, but more often from gross oversimplifications. Once the essential mechanism has been identified, it is easy to check for robustness by tagging on more and more details. It is usually easy to start at the simple and proceed to the complicated by adding more and more ingredients. On the other hand, it is an art to start at the complicated and messy and proceed to the simple and beautiful. The goal is not the reductionist one of identifying the "correct" underlying equations for evolution in all its details, but to set up some simple equations with the goal of illustrating robust processes. simplifications

Can We Model Darwin?

In the beginning of 1993 I had more or less accepted the failure of my frantic attempts to make Kauffman's NKC model and many other related models organize themselves to the critical state. Many trips to Santa Fe and numerous discussions had failed to lead to much progress.

This unhappy state of affairs changed suddenly when Kim Sneppen, a graduate student from the Niels Bohr Institute, came to visit us at Brookhaven for a week. Kim had started his career in nuclear physics, and had written scientific papers on fragmentation processes in heavy ion collisions. The Niels Bohr Institute has a glorious past in nuclear physics, sparked by Bohr's interest in the field. Bohr received his Nobel Prize for his quantum mechanical theory of the atom. That did not stop him from venturing into nuclear physics when that field opened up. However, many scientists at the Niels Bohr Institute have failed to realize that nuclear physics is not at the forefront of science any more, and not having Bohr's enthusiasm for new opportunities, they are stuck, living in a dream of past glory. Some of these older scientists even imitate Bohr's mannerism, such as his way of smoking a pipe. This has stifled the careers of two generations of physicists in Denmark who have seen the new horizons and are not content to live in the past. This is not so unusual; science is often driven by sheer inertia. Science progresses "death by death." A few young physicists have survived on temporary grants from Carlsberg and NOVO, two industrial giants in Denmark with vision and the willingness and ability to help out.

Kim had constructed a simple mathematical model for interfaces moving in a random medium. While superficially this might not seem more exciting than nuclear physics, at least it is different. Think, for instance, of coffee being absorbed by a paper napkin. The boundary between the wet paper and the dry paper forms an interface. The paper has some "pinning" sites where it is difficult for the interface to pass, as for instance narrow pores in the napkin (Figure 29). In his model, growth takes place at each time step at the site with the *smallest* value of the pinning force. The interface shifts upward by one length unit and is assigned a new random pinning force. This type of dynamics, where activity occurs at the place with the smallest or the largest value of some force, is called "extremal dynamics." Because of the elasticity of the interface, the growth at one site reduces the pinning force on the neighbor sites, making them likely candidates for growth at the next instance. Kim showed that the surface organizes itself to a critical state, with

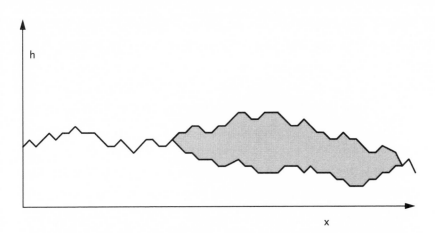

Figure 29.   Schematic picture of an avalanche separating two interface configurations in the Sneppen model. The size s of the avalanche, or punctuation, corresponds to the shaded area.

avalanches of all sizes. In other words, interface growth is a self-organized critical phenomenon. Similar ideas were developed earlier by Sergei Zaitsev at the Institute for Solid State Physics at the Chernogolovka research center near Moscow in a different context.

Kim gave a seminar at our physics department to a small audience of ten to fifteen scientists. He is not a polished communicator. His approach is spontaneous and intuitive, not based on detailed planning. His talk was like a dialogue with the audience with lots of questions back and forth. He is totally uninterested in his personal appearance, at least when he gets deeply involved in science. But his message came through.

I gave a brief spontaneous presentation on the evolution story at the blackboard in my office. In addition to Kim Sneppen, some of my coworkers at Brookhaven were jammed into the small office. Albert Libchaber, another visitor, famous for his work on chaos, was present. He had experimentally verified Feigenbaum's theory of the onset of chaos through period doubling bifurcations in a turbulent system. He shared with Feigenbaum the Wolf Prize, second only to the Nobel Prize in prestige, for that work.

"This is not a success story," I started, expressing my frustration. Then I related the story of the sandpiles, discussed Kauffman, Gould, and punctuated equilibrium, and ended with our futile studies of Kauffman's NKC models of rubber landscapes. There were a lot of comments from everyone. "I think that we can combine this with my way of thinking," Kim exclaimed at the end of the presentation.

The next day, a warm and nice spring Saturday, Kim and I went sightseeing on Long Island. We spent some time at a local fair that we happened to pass by; we saw a spectacular magic show and other exhibits. Then we went to Fire Island, a narrow island with miles and miles of beaches running parallel to Long Island along the south shore. On and off, we would discuss the evolution problem in a joking and playful way.

I don't know why it is, but it appears to me that this is the only way of doing imaginative science. The harder one tries, the less likely the prospect of success. I certainly never came up with any ideas by sitting intensely in my office staring at a sheet of blank paper.

A Science Project
for a Sunday Afternoon

On Sunday afternoon we went to work. *Extremal* turned out to be the magic word. Kim's model worked because the site with the "least" pinning force was selected for action. In fact, in the continuous deterministic sandpile models, which describe a bowl of sugar that is gradually tilted, avalanches start at the point with the largest slope. In earthquakes, the rupture starts at the location where the force first exceeds the threshold for breakage. Maybe extremal dynamics is the universal key to self-organized criticality. Could the principle be applied to models of evolution and thereby produce punctuated equilibria?

In the computer simulations that Stuart Kauffman and I had done, new coevolutionary avalanches were initiated by making a random mutation of a random species, that is by changing an arbitrary 1 to a 0 or vice versa somewhere in the NKC model. Kim and I decided to choose the species positioned at the *lowest* foothill in the Sewall Wright's fitness landscape for elimination, and replace it with a new species. Didn't Darwin invoke survival of the *fittest*, or, equivalently, elimination of the least fit?

One might think of this fundamental step either as a mutation of the least fit species, or the substitution of the species with another species in its ecological niche, which is defined by its coupling to the other species with which it interacts. Such an event is called a pseudo-extinction event. This is in line with Gould's picture of speciation: it takes place because of the "differential success of essentially static taxa." It is a matter of definition as to how many steps are needed to conclude that a species has become extinct and a new one has emerged, i.e., when a real extinction event has taken place. According to Sepkoski, "A species is what a reputable taxonomist defines as such." In our model, the number of species is conserved. Only the fitter of the original species and its mutated version is conserved.

The basic idea is that the species with the lowest fitness is the most likely to disappear or mutate at the next time step. These species (by definition) are most sensitive to random fluctuations of the climate and other external forces. Also, by inspection of the fitness landscape, it is obvious that in general the species sitting at the lowest fitness peak has the smallest valley to overcome in order to jump to a fitter peak. That is, the smallest number of coordinated mutations are needed to move to a better state. In fact, laboratory experiments on colonies of bacteria show that bacteria start mutating at a faster rate when their environment changes for the worse, for instance when their diet changes from sugar to starch.

However, we first wanted a simpler representation of the fitness landscapes than Stuart's cumbersome NKC landscapes. In the NKC models, a specific fitness was assigned to each combination of 1's and 0's in the genetic code. For a species with a twenty-bit code, interacting with four other species, we would have to store 2 to the 24th power random numbers, that is more than ten million numbers for each species. If there are 1,000 species, we would have a total of more than 10 billion numbers. In our model, we would not keep track of the underlying genetic code, but represent each species by a single fitness value, and update that value with every mutation of the species. We don't know the explicit connection between the configuration of the genetic code and the fitness anyhow, so why not represent the fitness with a random number, chosen every time there is a mutation? We then had to keep track of only 1,000 fitness values. If someone has the patience and computer power it ought to be possible to go back to an explicit representation of the fitness landscape.

Kim started to convert our ideas into computer language on my computer, an IBM workstation. We chose the species to be situated on the rim of a large circle. Each species is interacting with its two neighbors on the circle. This could represent something like a food chain, where each species has a predator on its left and a prey on its right. In principle, it could also have a symbiotic relationship with either. In the beginning of the simulation, we assigned a random number between 0 and 1 to each species. This number represents the overall fitness of the species, which can be thought of as positioned on a fitness peak with that random value of fitness. Then, the species with the lowest fitness was eliminated and replaced by another species. What would the fitness of the new species be? We tried several possibilities that worked equally well. The fitness of the new species after a mutation is unlikely to be much improved. One would not expect a jump from a very low peak to a very high peak. Thus, first we replaced the least fit species with a species with a fitness between 0 percent 10 percent higher than the original one. We also tried a version in which the new fitness is restricted to be between its old value and 1. However, for mathematical simplicity we finally tried to use a species with a completely random fitness. That means we assigned a new random number between 0 and 1 to that site. Of course, this does not represent real life. The important point is that the outcome of the simulation was robust with respect to these variations, so with a little bit of luck it might be broad enough to include real evolution.

The crucial step that drives evolution is the adaptation of the individual species to its present environment through mutation and selection of a fitter variant. Other interacting species form part of the environment. We could in principle have chosen to model evolution on a less coarse-grained scale by considering mutations at the individual level rather than on the level of species, but that would make the computations prohibitively difficult.

The idea that adaptation at the individual (or the species) level, is the source of complexity is not new. Zipf's observation that organization stems from the individuals' pursuit to "minimize their efforts" can be put in that category. In his book *Hidden Order,* John Holland, a computer scientist at the Santa Fe Institute and the University of Michigan, also locates the source of complexity to the adaptation of individuals. His observation is correct, but perhaps not particularly deep. Where else could complexity come from? Holland is best known for inventing "genetic algorithms" for problem solving. In these algorithms, the possible solutions to a given problem are represented by a genetic string of 1's and 0's, and the solutions evolve by random mutations and selection of the most fit variants, which is the variant that best solves the problem. The crucial issue is, again, *how* to go from the microscopic individual level to the higher level of many individuals where complexity occurs. We shall see that this happens because myriad successive individual adaptation events eventually drive the system of individuals into a global critical state.

How should we represent the interactions with other species? The reason for placing the species on a circle was to have a convenient way of representing who is interacting with whom. A given species would interact with its two neighbors, one to the left and one to the right. If the species that changes is the frog, the two neighbors could be the fly and the stork. We wanted to simulate the process by which the neighbors are pushed down from their peak and adjust by climbing the nearest peak available in the new landscape. One possibility was to choose the resulting fitness as some fixed amount, say 50 percent lower than the original peak. We tried this and many different algorithms for choosing the new fitness of the neighbor. The programs were so simple that the programming for each version would take no more than ten minutes, and the computer run would take only a few seconds to arrive at some rough results. Again, the interactions could be chosen arbitrarily, which is crucial since without this type of robustness the model could not possibly have anything to do with real evolution. We settled on a version where the fitnesses of the neighbors would simply change to new random numbers between o and 1.

In summary, the model was probably simpler than any model that anybody had ever written for anything: *Random numbers are arranged in a circle. At each time step, the lowest number, and the numbers at its two neighbors, are each replaced by new random numbers.* That's all! This step is repeated again and again. What could be simpler than replacing some random numbers with some other random numbers? Who says that complexity cannot be simple? This simple scheme leads to rich behavior beyond what we could imagine. The complexity of its behavior sharply contrasts with its simple definition.

In a business context, the process would correspond to a manager firing the least efficient worker and his two coworkers, and then replacing them with three new guys coming in from the street. The abilities that the two coworkers had learned by working with their poor performing colleague would be useless. Of course, the manager's rule is not fair, but neither are the laws of nature.

At the start of the computer simulation, the fitnesses on average grow since the least fit species are always being eliminated. Figure 30 shows the fitness of the least fit species versus time. Although there are fluctuations up and down, there is a general tendency of the average fitness to increase. Eventually, the fitnesses do not grow any further on average. All species have fitnesses above some threshold. The threshold appears to be very close to $2/3$. No species with fitness higher than this threshold will ever be selected for spontaneous mutation; they will never have the lowest fitness. However, their fate may change if their weak neighbors mutate.

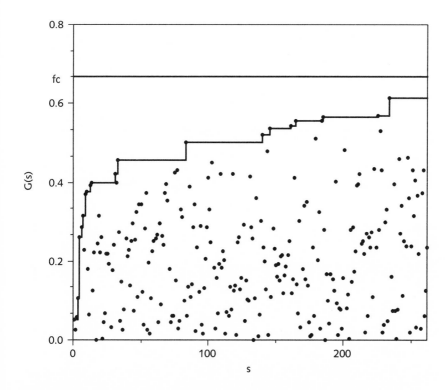

Figure 30.   Fitness of least fit species vs. the number *s* of update steps in a small evolution model with twenty species. The envelope function, defining the fitness gap, increases in a stepwise manner. An avalanche starts when there is a step, and ends at the next step, where a new avalanche starts. The envelope function eventually reaches the critical value f_c (Paczuski et al., 1995).

Let us consider a point in time when all species are over the threshold. At the next step the least fit species, which would be right at that threshold, would be selected, starting an "avalanche," or "cascade," or "punctuation" of mutation (or extinction) events that would be causally connected with this triggering event. There is a domino effect in the ecology. After a while, the avalanche would stop, when all the species are in the state of "stasis" where all the fitnesses again will be over that threshold.

Figure 31 shows a snapshot of all the fitnesses of all the species in the midst of an avalanche in an ecology consisting of 300 species. Note that most species are above the threshold but there is a localized burst of very active

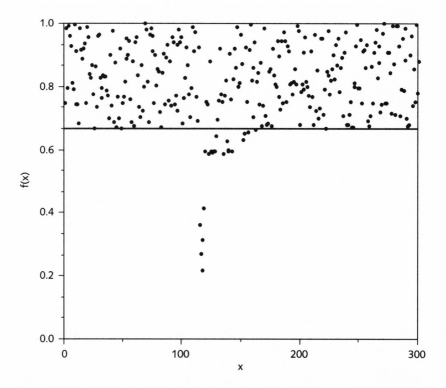

Figure 31.   Snapshot of the fitnesses in the stationary critical state in the evolution model. Except for a localized region where there is a relatively small fitnesses due to a propagating coevolutionary avalanche, all the fitnesses in the system have fitnesses above the self-organized threshold $f = 0.6670$ (Paczuski et al., 1995).

species with fitnesses below the threshold. Those species will be selected for mutation again and again, as the avalanche moves back and forth in the ecology. The species with high fitnesses are having a happy life, until the avalanche comes nearby, and destroys their pleasant existence. In some sense, nature is experimenting with all kinds of mutations, until it arrives at a stable complex network of interacting species, where everybody is stable, with fitnesses above the threshold. One can think of this as a learning process in which nature creates a network of functionally integrated species, by self-organization rather than by design. The "blind watchmaker" is at work. The Cambrian explosion 500 million years ago, and the Permian extinction 250 million years ago in which 96% of all species became extinct, were the biggest avalanches that have occurred so far. At the Cambrian explosion, beautifully described in Stephen Jay Gould's book *Wonderful Life,* nature experimented with many different designs, most of which were discarded soon after, but out of the Cambrian explosion came a sustainable network of species.

We observed a similar behavior in the Game of Life. An avalanche of unstable, low-fit individuals with short life spans propagates until the seemingly accidental emergence of a stable network of organisms.

We monitored the duration of the avalanche, that is the total number of mutation events in each avalanche, and made a histogram of how many avalanches of each size were observed. We found the all-important power law. There were indeed avalanches of all sizes, $N(s) = s^{-\tau}$, with τ being approximately equal to 1. Small avalanches and large avalanches are caused by the same mechanism. It does not make sense to distinguish between background extinctions happening all the time, and major ecological catastrophes.

That afternoon, we simulated five or six versions of the model, and the result was always the same, with the same value of the exponent τ. In that sense, our result appeared to be universal. The system had self-organized to the critical state.

For a change, we left the lab with a great sense of accomplishment that evening. It was no longer a fundamental mystery to us how an interacting ecology could evolve to a "punctuated equilibrium" state with ecological catastrophes of all sizes. Of course one might want to put some more meat on the skeleton of the model that we had constructed, but we were confident that the fundamental conclusion would survive. Darwin's mechanism of selecting the fitter variant in an ecology of species leads not to a gradually changing ecology but to an ecology in which changes take place in terms of coevolutionary avalanches, or punctuations. Our numerical simulations had demonstrated that there is no contradiction between Darwin's theory and punctuated equilibria. Our model is in the spirit of Darwin's theory, but nevertheless exhibits punctuated equilibria.

The effort that afternoon is an example of a working model with interactions between man and machine that could not have taken place even a decade ago. The efficiency and availability of small computers have reached the point where one can obtain answers as soon as one can think up simple models. A few years ago, a project like this would have taken weeks, punching cards and waiting for output at a central computer, rather than the afternoon the project actually consumed. It would have been utterly impossible for R. A. Fisher and his contemporaries to do something similar in the 1930s.

Let us briefly return to what went wrong in our previous attempts to model punctuated equilibria. First of all, the idea that the critical point represents a particularly "fit" or good state was misguided. When we see ourselves and other species as "fit" this means that we are in a period of stasis in which we form a stable, integrated part of a complex ecological network. Let it be cooperation or competition. The key point is that the network is self-consistent, just like Conway's creatures in the Game of Life.

We are "fit" only as long as the network exists in its current form. We tend to see fitness as something absolute, perhaps because we view the present period of stasis as permanent, with a preferential status. However, when the period of stasis is over, it is a new ball game and our high fitness might be destroyed. Actually, in a greater perspective, our present period might not even be a major period of stasis, but a part of an avalanche. Life is unstable and volatile. Dead, inert material is stable and in this sense fit. Ironically, evolution cannot be seen as a drive toward more and more fit species, despite the fact that each of the steps that constitutes evolution may improve the fitness.

What one species (humans) may see as its superior fitness may better be characterized as a self-consistent integration into a complex system. Seen in isolation, the emergence of organisms as complicated as us is a complete mystery. Biology constructed the solution to the fitness problem together with the problem itself by a process of self-organization involving billions of species. It is a much simpler task to construct a complicated crossword puzzle by a coevolutionary process than to solve it by trial and error for each word in isolation. Evolution is a collective Red Queen phenomenon where we all keep running without getting anywhere.

Our simple model barely constitutes a skeleton on which to construct a theory of macroevolution. It is not the last word on the matter; probably it is the first. It is a simple toy model that demonstrates how, in principle, complexity in an interacting biology can arise. It is the beginning of a new way of thinking, not the end. It ignores an embarrassing range of real phenomena in evolution. There is no process by which the number of species can change. Why are there species in the first place? Also, the fitness landscape is introduced ad hoc. In a more realistic theory, the landscape itself should be self-organized in the evolutionary process. However, we believe that our model is a useful starting point for these considerations. Indeed, there has been a flood of activity; scientists have augmented our model to make it more complete. Vandevalle and Ausloos of Liege, Belgium, have included speciation. The mutating species gives rise to two or three new species, each with its own fitness, which enter the ecology in competition with all the other species. Vandevalle and Ausloos start their simulation with a single species. This results in phylogenetic tree structures, with a hierarchical organization similar to the taxonomic classification of species into phyla, genera, and families. The model still self-organizes to the critical state. The exponents of the power law are different from ours.

Why is it that the concept of punctuated equilibrium is so important for our understanding of nature? Maybe the phenomenon illustrates better than anything else the criticality of a complex system. Systems with punctuated equilibria combine features of frozen, ordered systems, with those of chaotic, disordered systems. Systems can remember the past because of the long periods of stasis allowing them to preserve what they have learned through history, mimicking the behavior of frozen systems; they can evolve because of the intermittent bursts of activity.

Life at a Cold Place

In real life there is no Grim Reaper looking for the least stable species, asking it to put up (mutate) or shut up (go extinct). Things must happen in parallel everywhere. A real-time scale for the mutations has to be introduced. Species with low fitness, at the low peaks in the fitness landscape, have a short time scale for jumping to better maxima; species with high fitness are less inclined to mutate because a larger valley has to be traversed to find a more fit peak.

The barrier that has to be traversed can be thought of as the number of coordinated mutations of the DNA that have to occur to take the species from one maximum to a better one. The number of random mutations that have to be tried out increases exponentially with this barrier. Thus, the time scale for crossing the barrier is roughly exponential in the fitness. One can think of the probability of a single mutation as given by an effective temperature T. For high temperatures, there is a high mutation rate everywhere, and the dynamics are very different from the punctuated equilibrium behavior discussed here. There cannot be large periods of stasis in systems that are disturbed at a high rate. If the sandpile is shaken vigorously all the time it cannot evolve to the complex, critical state. It will be flat instead. For low temperatures, or for low mutation activity, the dynamics studied here are recovered without explicitly searching for the species with the lowest fitness.

We arrive at the conclusion that complex life can only emerge at a cold place in the universe, with little chemical activity—not a hot sizzling primordial soup with a lot of activity.

Comparison with Real Data

To arrive at an overview of evolution in the model, one can make a space time plot of the evolutionary activity (Figure 32). The x axis is the species axis, and the y axis is time. The plot starts at an arbitrary time after the self-organized critical state has been reached. A black dot indicates a time that a given species undergoes a mutation. The resulting graph is a fractal. Starting from a single mutating species, the number R of species that will in average be affected after a large number S of updates will be a power law, $S = R^D$, where the exponent D is called the "fractal dimension" of the graph.

To monitor the fate of individual species, let us focus on a single species, for instance the one situated on the origin of the species axis, as we move along the vertical time direction. Obviously, there are long periods with no black dots when not much is going on. These are the periods of stasis. Also, there are some points in time when there is a lot of activity. Let us count the number of mutation events as we move along the time direction. Figure 33 shows the accumulated number of mutations of the selected species as a function of the time. One can think of this number as representing the amount of physical change, such as the size of a horse, versus time. The "punctuated equilibrium" nature of the curve is obvious. There are long periods of stasis where there is no activity, separated by bursts of activity. Such a curve is called a Devil's staircase because of

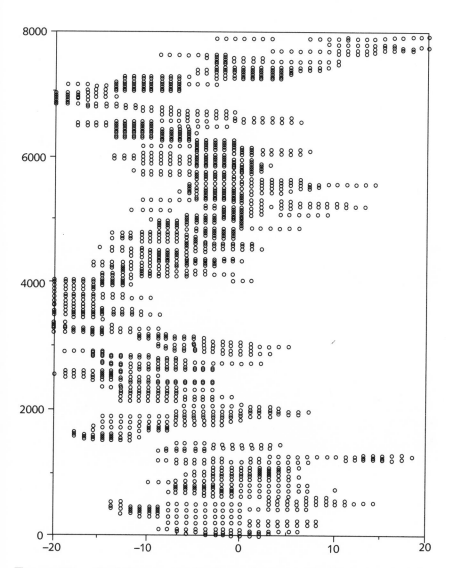

Figure 32.   Activity pattern for the evolution model. For each species, the points in time where it undergoes a mutation is shown as a black dot. Time is measured as the number of update steps. The pattern is a fractal in time and space (Maslov et al., 1994).

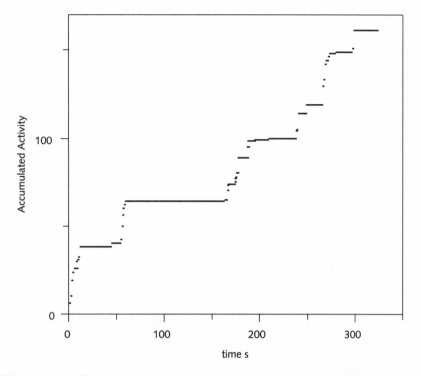

Figure 33.   Punctuated equilibria in the evolution model. The curve shows the number of mutation events for a single species, that is the number of black spots encountered when moving along the vertical direction through the fractal shown in Figure 32.

its many steps, some very large, but most very small. Between any two steps, there are infinitely more steps. The Devil's staircase was invented by the German mathematician Georg Cantor (1845–1918) in the nineteenth century, and for a long time it was thought that no physical system could possibly show such intricate behavior.

One can measure the distribution of the durations of the periods of stasis, or the return times between mutations. There are no real jumps in the curve, only periods with a large number of very rapid small increases. In the fossil record, one might not be able to resolve these small, rapid increases, so the resulting variation appears as a jump, or saltation. For comparison, Figure 34

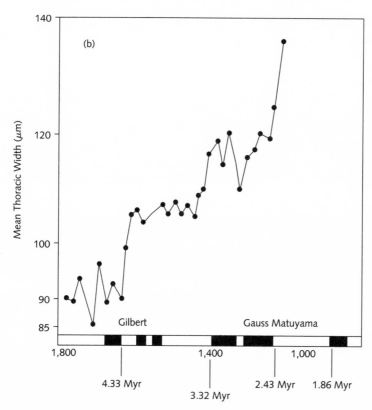

Figure 34. Punctuated equilibrium in nature. Thoracic width of the radiolarian *Pseudocubus vema* as it has increased through its evolutionary history (Kellogg, 1975).

shows how the thoracic width of the radiolarian *Pseudocubus vema* has evolved during the last five million years. This curve has a quite similar structure to the one in Figure 33. Note that there are no large jumps in the curve. The punctuations are simply periods where there is a large amount of evolutionary activity. The evolution of the size of the horse follows a similar pattern.

In our crude model, the single step can be thought of as representing either an extinction event, in which the niche of the species that became extinct is filled by another species, or a pseudo-extinction event, in which a species mutates into a different species. In either case, the original species does not exist after the event. In real evolution the same question may arise. Species may become extinct, or they may mutate through several steps into something quite different. The statistical properties of avalanches in our model should be similar to the statistical properties of extinction events in biological history. Therefore, it makes sense to compare the results of simulations with the record of extinction events in the fossil record.

By running the computer long enough, we can accumulate enough data to make the statistics of our model very accurate. In one run, we made more than 400,000,000,000 pseudo-extinctions. That is more than eighty mutations for each person in the world. We can also make several runs on the computer, whereas there is only one evolution of history on earth. It is impossible for even very meticulous paleontologists like John Sepkoski to compete with this, making it difficult to compare our predictions with reality. Sepkoski looked at "only" 19,000 real extinctions of species.

To make comparisons with data, Kim Sneppen, Henrik Flyvbjerg, Mogens Jensen, and I have simulated the evolution model in real time units as discussed above. We sampled the rate of extinctions (or pseudo-existence) taking place in temporal windows of a few hundred time steps, to be compared with Sepkoski's binning of data in intervals of four million years. In this way we were able to generate a synthetic record of extinctions (Figure 35). Note the similarity with Sepkoski's data (Figure 4).

Raup's histogram of Sepkoski's data in Figure 5 can be reasonably well fitted to a power law with exponent between 1 and 3. Figure 36 shows, for comparison, the distribution of extinction events from the model. The important point is that the histogram is a smooth curve with no off-scale peaks for large extinction events. It would certainly be nice to have a finer resolution on the data, with extinctions measured, say, every one million years.

Sepkoski also noted that extinctions within individual families were correlated with extinctions in other families across the various taxa. One may say that the evolutions of different species "march to the same drummer." This is exactly what to expect from our simulation, in which extinction events, in-

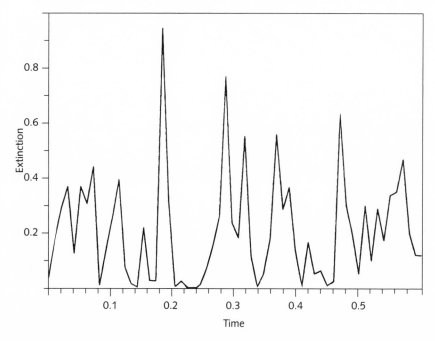

Figure 35. Synthetic record of extinction events from the punctuated equilibrium model. Note the similarity with Sepkoski's curve for real evolution (Figure 4).

cluding mass extinctions, can be thought of as the radiation of adaptive changes of individual species.

Figure 37 shows the accumulated mutations of a single species, the Devil's staircase, together with a plot of the global activity of extinctions. A real time scale in which the mutations rate was represented by a low temperature was used. The individual species change during periods when there is a large general activity, as observed by Sepkoski, although not all avalanches affect the species that we are monitoring. No outside "drummer" is necessary, however. The synchronized extinctions are a consequence of the criticality of the global ecology, linking the fates of the various species together, like the sand grains of the sandpile model.

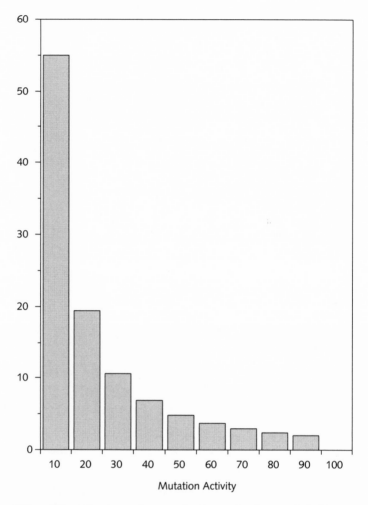

Figure 36.   Distribution of events in the evolution model. Compare with Raup's plot (Figure 5). The events can be thought of better as extinctions, or pseudo-extinctions where a species disappears by mutating into another species.

Although large events occur with a well-defined frequency, they are not periodic, neither in real evolution nor in our simulation. For real evolution, this has been pointed out repeatedly, most recently by Benton in his book *The Fossil Record*. The actual statistical properties of the extinction record supports the view that biological evolution is a self-organized critical phenomenon.

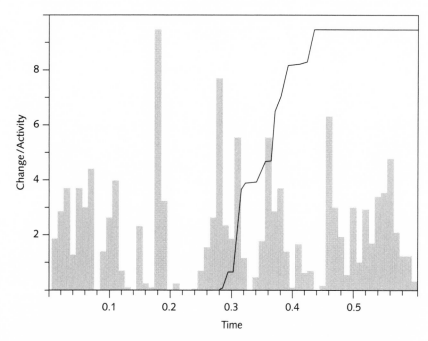

Figure 37.   Global vs. local activity in the evolution model. The global extinction rate is indicated by the gray histogram. The curve shows the evolution of a single, randomly chosen species. The bursts of rapid activity take place during periods of large extinction activity. Evolution of different species "march to the same drummer." This was noted by John Sepkoski for real evolution (Sneppen et al., 1995).

On Dinosaurs and Meteors

Implicit in all proposed causes of mass extinction so far, including the theory involving an asteroid impact, is a presumed equality between cause and effect. According to this philosophy, mass extinction must be caused by a cataclysmic external event, and the only way to understand the extinction event is to identify that event. Alvarez' theory of a meteorite hitting the earth sixty million years ago, and thereby causing the extinctions of dinosaurs is widely accepted. Alvarez even suggested that the meteorite was one falling at his own doorstep, near the Yucatan peninsula in Mexico. The remnants of a large crater and a layer of iridium spreading worldwide at about the same time are taken as evidence of the theory. One reason may be that no "alternative" theory has emerged, in the sense that no other cataclysmic impact has been suggested.

The impact theory has been accepted despite two major shortcomings. First, the dinosaurs appear to have died out at least a couple of million of years before the meteorite hit. At the very least, the dominance of the dinosaurs was already greatly reduced at that time. It defies logic to claim that a meteor hitting when the dinosaurs were on the way out was responsible for their demise. There would be no obvious need for the meteor. The real question would be why the dinosaurs were going downhill in the first place. Second, no causal relationship between the meteor and the resulting extinction has been established. What actually killed the dinosaurs? All we have are loose, unsubstantiated, speculations about climate changes caused by the meteor. And why were the dinosaurs affected and not certain other species?

The fact that extinctions are synchronized is taken as further evidence, in particular by Niles Eldridge, of an external force working across families. Indeed, in an equilibrium linear world there would be no other possibility. A massive extinction event requires a massive external impact. This is not the case in our self-organized critical world.

Our calculation demonstrates that it is at least conceivable that the intermittent behavior of evolution, with large mass extinctions, can be due to the internal dynamics of biology. In his book *Extinction: Bad Luck or Bad Genes?*, Raup argues that extinction is caused by bad luck due to external effects, rather than by intrinsically bad genes. We argue that even in the absence of external events, good genes during periods of stasis are no guarantee of survival. Extinctions may take place also due to bad luck from freak evolutionary events endogenous to the ecology. This cannot rule out that extinction events were directly caused by some external object hitting the earth. Of course, in the greater picture, nothing is external so in the final analysis catastrophes must be explainable endogenously in any cosmological model.

However, the fact that the histogram of extinction events is a smooth curve indicates that the same mechanism is responsible for small and large extinction events, because otherwise the size and frequency of these large events would have no correlations with the smaller extinction events. Certainly the extinctions taking place all the time have nothing to do with extraterrestrial impacts.

In fact it is quite simple and natural to reconcile the two viewpoints. In our model the avalanches were initiated by events that we thought of as mutations of a single species. One might also think of the initiating event as having an external cause. Think of the sandpile model in which the avalanches are initiated by dumping a grain of sand from the outside. Within this latter interpretation, the meteor hitting the earth merely represents a triggering event, which initially would affect only a single or a few species. Maybe it destroyed some vegetation because of lack of sunshine. The demise of these species would destroy the livelihood of other species, and so on. The resulting mass extinction would be a domino process "caused" by this initial event. The mass extinction could only take place because the stage had been set by the previous evolutionary history, preparing the global ecology in the critical state. In a couple of recent articles, Newmann has extended our model to include the effect of external perturbations as sketched here. They still find self-organized criticality (SOC) with a power law distribution of avalanches, although their value of the exponent, $\tau = 2$, is different from ours and possibly in better agreement with Raup's and Sepkoski's observations.

Dante Chialvo's Evolutionary Game

Dante Chialvo is a colorful Argentine, originally trained as a medical doctor, now living in the United States. He has given up his original career and performs research, mostly theoretical, on brain modeling and evolution. I came to know him at a conference on self-organized criticality, stochastic resonance, and brain modeling that he organized in 1990, in Syracuse, New York. It was not clear at all at that point what SOC has to do with brain modeling. I guess that the conference was organized in an attempt to connect medical brain research with current ideas in dynamic systems. After all, the brain *is* a large dynamic system with myriad connected neurons—we shall return to this later.

That meeting brought me into contact with a group of scientists looking for general mechanisms for the organization of living organisms, covering an enormous spectrum of thoughts. The signal-to-noise ratio of the talks and the discussions was rather low, but at least here was a group of open-minded people realizing a need for new ideas.

From Syracuse, Dante moved to the Santa Fe Institute for a couple of years, and then on to the University of Arizona to take a teaching position. Having heard about our evolution model at various conferences and at bar conversations, Dante came up with his own pedagogical version.

Dante arranged his twenty students in a circle and gave them twenty-sided dice. The students represented the species, and the number on his die is his fitness. Our random number generator is replaced by a throw of the dice. At each step the student with the lowest number, that is the species with the least fitness is selected. He throws his die, and so do his two neighbors. The new random numbers represents their new fitnesses. In case two guys share the lowest number, the one to go extinct would be decided by a roll of a tie-breaker die. The student who now would hold the lowest number is then selected for extinction and so on. A twenty-first student would do the bookkeeping at the blackboard. He would monitor and plot the running smallest number of all the dice. That would trace out a curve looking like Figure 30.

After several rolls, most of the students would be looking at numbers exceeding a critical fitness threshold of 13, that is near the fraction 0.667 found in our model. The bookkeeper then starts measuring the avalanche distribution. An avalanche starts when the lowest number among all the students exceeds 13, and it stops when the lowest number exceeds 13 again. The whole dynamics can be followed in detail. Because of the small number of students and their limited patience, the resulting statistics are lousy compared with what can be obtained from the high-speed digital computer. Punctuated equilibrium behavior can be detected by plotting the accumulated activity of a single, selected student. If we count how many times he has thrown his die up to a time t, the resulting curve will look somewhat like Figure 33. For long periods of time, the periods of stasis, he does not throw the die at all, while other students are busy, but this inactivity is interrupted by relatively short periods where he and his neighbors get busy.

Self-Organized Criticality and Gaia

In a seminal work, Jim Lovelock, an English scientist, came up with the fascinating idea that all life on earth can be viewed as a single organism. This idea has struck many scientists as preposterous since it flies in the face of the usual reductionist approach and smacks of New-Age philosophy. Lovelock's idea is that the environment, including the air that we breathe, should not be viewed as an external effect independent of biology, but that it is endogenous to biology. The oxygen represents one way for species to interact. Lovelock noted that the composition of oxygen has increased dramatically since life originated. The oxygen content is far out of equilibrium. The layer of ozone, an oxygen molecule, that protects life on earth did not just happen to be there, but was formed by the oxygen created by life itself. Therefore, it does not make sense to view the environment, exemplified by the amount of oxygen, as separate from biological life. One should think of the earth as one single system.

What does it mean to say that the earth is one living organism? One might ask in general: What does it mean that anything, such as a human, is one organism? An organism may be defined as a collection of cells or other entities that are coupled to each other, so that they may exist, and cease to exist, at the same time—that is, they share one another's fate. The definition of what represents an organism depends on the time scale that we set. In a time scale of 100 million years, all humans represent one organism. At short time scales, an ant's nest is an organism. There is no fundamental difference between genetically identical ants carrying material back and forth to build and operate their nest, and genetically identical human cells organizing themselves in structures and sending blood around in the system to build and operate a human body.

Thus a single organism is a structure in which the various parts are interconnected, or "functionally integrated" so that the failure of one part may cause the rest of the structure to die, too. The sandpile is an organism because sand grains toppling anywhere may cause toppling of grains anywhere in the pile.

One might think of self-organized criticality as the general, underlying theory for the Gaia hypothesis. In the critical state the collection of species represents a single coherent organism following its own evolutionary dynamics. A single triggering event can cause an arbitrarily large fraction of the ecological network to collapse, and eventually be replaced by a new stable ecological network. This would be a "mutated" global organism. At the critical point all species influence each other. In this state they act collectively as a single meta-organism, many sharing the same fate. This is highlighted by the very existence of large-scale extinctions. A meteorite might have directly impacted a small part of the organism, but a large fraction of the organism eventually died as a result.

Within the SOC picture, the entire ecology has evolved into the critical state. It makes no sense to view the evolution of individual species independently. Atmospheric oxygen might be thought of as the bloodstream connecting the various parts of our Gaia organism, but one can envision organisms that interact in different ways.

The vigorous opposition to the Gaia hypothesis, which represents a genuine holistic view of life, represents the frustration of a science seeking to maintain its reductionist view of biological evolution.

Replaying the Tape of Evolution

In real life we cannot "rewind the tape of evolution," but in our simple model we can! History and biological evolution are massively contingent on spurious incidents. The question of what if this or that did not happen has been the source of endless speculations by historians, and has constituted material for numerous books and movies. What if Lee Harvey Oswald had missed John F. Kennedy in Dallas? Would world history have changed? What if Columbus had been forced to return, or hit a hurricane on his dangerous journey into the unknown? In the movie *Back to the Future,* McFly returns to the past and changes a few minor details, thus repairing some bad features of his present life. In the current television show "Gliders," a group of travelers visit the earth in various parallel universes. In one episode one has to stop for green light instead of red; in another the Russians won the Cold War and transformed Alaska into Gulag Archipelagos. In real life, we never know what would have happened. We cannot extrapolate from our present situation into the future (or from the past into the present). Where will the stock market be in a year from now? Or tomorrow?

One could argue that it is actually the sensitivity of real life to minor spurious events that makes fiction possible, or believable. One could not think of literature, apart from the most boring, describing life in a noncritical universe where everything is ordered and predictable. That world could not be subjected to realistic and believable manipulations by the fiction writer. Nor can one have a literature in a world where everything is totally random and chaotic, because then what happens tomorrow has nothing to do with what happens today.

The importance of contingency in economics has been stressed by Brian Arthur of the Santa Fe Institute. As an example, he argues that the victory of the VHS system over Betamax for video recording, or combustion engines over steam engines, was dependent on spurious historical events rather than on the technical superiority of the winning project. In traditional equilibrium economics the best product always wins.

Stephen Jay Gould has emphasized the role of contingency in determining the history of life on earth. One of my colleagues, Maya Paczuski, had been reading Gould's books and mentioned that the importance of contingency could be understood as a consequence of self-organized criticality. What if we were actually able to replay history under slightly different circumstances? In real life, everything occurs only once in its full glory, so we can't do that. But in our simple model of evolution we can play God and perform the computer simulation again, with only a tiny modification somewhere.

How could we make this idea concrete? We decided to "rewind the tape of evolution." At first we ran the evolution model as usual and monitored the accumulated number of mutations at one site (Figure 38), recovering the usual punctuated equilibrium Devil's staircase. We then identified the event that initiated one of the larger avalanches involving that particular site. Of course, that could be done only in hindsight. This event happened to be at a distance from the particular species that we monitored. We eliminated that event by replacing the fitness with a higher value and thus preventing extinction there. This interruption could correspond to changing the path of a meteor, or preventing the frog from developing its slippery tongue. We then ran the simulation again. The random numbers that were chosen were the same as before for species not affected by the small change that we made. New random fitnesses were chosen whenever needed for species that were affected by the change, and any future event that was affected. At the point where the minor perturbation

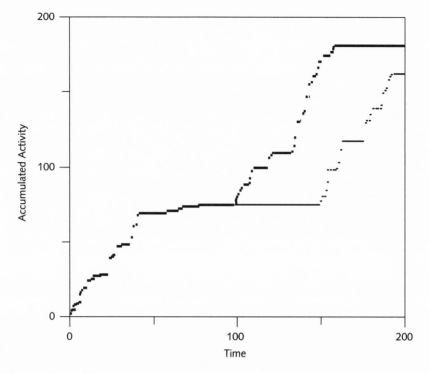

Figure 38.   Replaying the tape of history. After running the evolution model once (fat curve) the evolution was run once more, with a single random number changed so that a mutation event was eliminated (broken curve). A large catastrophic extinction event was avoided, but others occurred later in the evolutionary history (Bak and Paczuski, 1995).

was made, history changed. The accumulated number of mutation events in the replay of evolution was monitored as in the original history.

The new result is shown as the broken curve in the figure. The large punctuation is gone. However, that did not prevent disasters at all. Other punctuations happened at later points. Thus large fluctuations cannot be prevented by local manipulation in an attempt to remove the source of the catastrophe. If the dinosaurs had not been eradicated by a meteor (if they indeed were), some other large group of species would be eliminated by some other triggering event.

Because of the large sensitivity of the critical state, a small perturbation will eventually affect the behavior everywhere. Chaos scientists call this the *butterfly effect.* A butterfly moving its wings in South America will affect the weather in the United States. What they have in mind is a simple system like the Feigenbaum map, or a pushed pendulum, or small number of coupled differential equations. If one gives the pendulum a microscopic extra push, the position of the pendulum at later times will greatly differ from the original trajectory in an unpredictable way. Of course, the global weather is not a simple chaotic system, so these considerations appear irrelevant. Our evolution model illustrates the butterfly effect for a complex system. Any small change of any event will sooner or later affect *everything* in the system. If the initial event caused a large avalanche, the effect will take place sooner rather than later. We believe that the effect that we have described is the real butterfly effect, in contrast to the one found in simple chaotic systems that have no relevance to evolution or any other complex system.

To illustrate the connection between criticality and punctuated equilibria, we also ran a simulation for a noncritical system. We stopped evolution *before* it had evolved to the critical point, and did the same two computer runs, with and without eliminating an extinction event. The noncritical evolution is gradual, with no large intermittent bursts. Changing or eliminating one roll of a die does not have any dramatic outcome whatsoever. In particular, species that are distant from the event that was eliminated were not affected at all in the simulation. All of these simulations can conveniently be done by means of Dante Chialvaco's dice game.

theory of the punctuated equilibrium model

The reader who is not mathematically and analytically inclined may skip most of this chapter, in which we take a brief look into the mathematical analytical theory of the punctuated equilibrium model, except for the final section, which points out an insightful analogy between evolution and earthquakes. It is important not to skip this section because the main point of this book is to prepare the ground for, and to develop, relevant analytical insight into the behavior of the model, and hence into the underlying physical processes. The main reason for dealing with grossly oversimplified toy models is that we can study them not only with computer simulations but also with mathematical methods. This puts our results on a firmer ground, so that we are not confined to general grandiose, philosophical claims.

As a fringe benefit, the insight achieved from the study of the simple evolution model can be applied to the Game of Life. providing a spectacular, totally unexpected link between theory on the most microscopic level—particle theory— and the complex behavior of Conway's Game of Life.

What Is a Theory?

Curiously, the concept of what constitutes a theory appears to be different in biology and physics. In biology, Darwin's thoughts about evolution are always referred to as a theory, even though it is only a verbal characterization of some general observations. There is nothing wrong with that. According to one of the most fundamental principles of science, a theory is a statement about some phenomenon in nature that in principle can be confronted with reality and possibly falsified. The description can be either verbal or mathematical. In physics, we use the language of mathematics to express our theories. To confront the theory with reality, we solve equations and compare with experiments. The result of the theory is a number that is compared with a number measured by some apparatus. If there is disagreement, we return to the drawing board. When theories are expressed verbally in terms of much less precise languages, the confrontation with facts is much more cumbersome and leaves space for endless discussions among experts as to what constitutes the better description. Sometimes the experimental observation itself, without any condensation into more general principles, is viewed as a theory.

The science of paleontology is an empirical observational science like astronomy and experimental particle physics. However, there seems to be a belief, based on some misguided inferiority complex acknowledged and discussed at great length in the paleontologist Stephen Jay Gould's *Wonderful Life,* that the science becomes more respectable if the word *theory* can be attached to it. The science has been dismissed as "theoretical impotent."

This ambiguity about what counts as a theory became clear to me at my first meeting with Gould. I was giving a physics colloquium at Harvard's physics department in 1993, just about the time when the original work with Kim Sneppen was completed. My host was David Nelson, professor of condensed matter physics. I expressed a wish to discuss our ideas with Gould, who is also a professor at Harvard. Unfortunately, my schedule for that day (not to speak of his) was completely full so nothing was arranged.

In the evening, David invited me to the Harvard Society of Fellows for dinner. There was barely time for that, since I had to take an 8 o'clock flight back from Boston to Long Island. I happened to be sitting next to the president of the society, and on the other side of the president was a smiling gentleman. I introduced myself. "Stephen Jay Gould," the gentleman responded. What a coincidence—the very person I wanted to meet was my neighbor at the table. That should not be wasted.

"Wouldn't it be nice if there were a theory of punctuated equilibria?" I started.

"Punctuated equilibria is a theory!" Gould responded.

Where do you go from there? Not much communication took place, and I had to run to catch my plane.

The Random Neighbor Version of the Evolution Model

How does one go about constructing a theory in the physicist's sense? The construction of a simple model in conjunction with computer simulations does not in itself constitute a full-fledged theory. Although the numerical results do provide predictions to compare with observations, they give only a limited amount of insight into the physical process of self-organized criticality. The main advantage of having simple models of complex phenomena is that one might eventually be able to deal with them with powerful mathematical methods. For that reason, we have stripped down the evolution model as far as possible. The computer simulations act as a guideline for the analytical approach. They help us focus our ideas. The model and the numerical simulation serve as a bridge between nature and a mathematical theory. The main theoretical issues to be addressed are the process by which the model organizes itself to the critical state, and the characterization of the critical state, expressed for instance in terms of the critical exponents for the power laws characterizing the critical state, which eventually should be compared with observations.

After constructing our model, and doing the first preliminary computer studies, we approached our colleague Henrik Flyvbjerg who has a more mathematical mind, and already had worked on Stuart Kauffman's models with Bernard Derrida of Saclay, France. Also, Henrik was the primary intellectual capacity in our theoretical proof that Kauffman's NKC models do not exhibit self-organized criticality.

Henrik was visiting Princeton University, so we agreed to meet midway, in Manhattan. While walking down Eighth Avenue from midtown all the way to Battery Park, Kim Sneppen and I explained to Henrik how we had finally come up with a self-organized critical model of evolution. It didn't take Henrik long to come up with a version that would yield to rigorous analysis. While we were having lunch, he also figured out a rigorous way of properly defining the avalanches in terms of the activity below the critical threshold in Figure 31.

Instead of placing the species in a circle, he let each species interact with two randomly selected species in the system. At each time step one would select the species with the lowest fitness, and two other random species, and provide all three with new random fitnesses. In Dante Chialvo's game version, that would correspond to a situation in which the student with the lowest value on the die and two other random students in the class would roll their dice at each time step.

Henrik calculated the fitness threshold above which all species would find themselves after a transient time. The threshold is at $1/3$, to be compared with 0.667 for the chain model. This number in itself is of no importance. He also calculated the exponent of the power law for the avalanche distribution, $\tau = 3/2$. There would be slightly fewer catastrophic events than in the original model in which τ was 1.07. This exponent seems to be in better agreement with Raup's data for the distribution of extinction events (Figure 5). Many other results are now available on the random neighbor model. As usual, the resulting mathematics turned out to be highly complicated despite the simple nature of the model.

The avalanche process in the random neighbor model can be thought of as a "random walk." At a given stage of the propagating avalanche, there will be a number of active species with fitnesses below the threshold. At the next time step, the number of active species will take a random step: the number will either increase or decrease by 1. The process continues until there are no more active species, and the avalanche is over.

In his book *Extinction: Bad Luck or Bad Genes?*, Raup has made some similar observations. Estimating the lifetimes of various families of species using Sepkoski's data, he suggested that the process indeed is a random walk, in which at each step the number of species in the family increases or decreases by 1. Unfortunately, Raup is not a good mathematician so his analysis of the consequences of that picture is flawed; he thought that it would lead to a "characteristic lifetime" of a couple of millions of years, in contrast to the power law without characteristic lifetimes of species. Henrik Flyvbjerg, Kim Sneppen, and I have analyzed Raup's "kill curve" (Figure 39) on which he based his theory, and realized that it is a very beautiful power law, with exponent 2. This might be one of the best indications that life is indeed a self-organized critical phenomenon. We do not understand why the exponent is 2.

There is another solvable model with a good deal more complexity. In 1993 to 1995 Stefan Boettcher was a research associate at Brookhaven working mainly in the theory of particle physics. He became interested in the world of self-organized criticality, following a general tendency of particle physicists to look elsewhere into less crowded areas of science. Maya Paczuski and

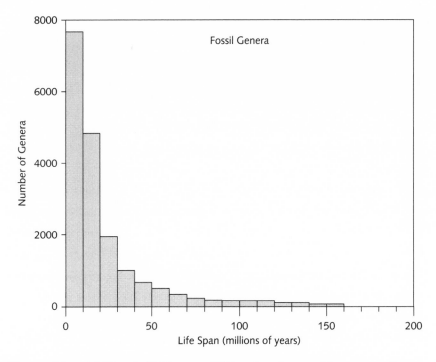

Figure 39.   (a) Raup's "kill curve." The plot shows a histogram of the number of genera with a given lifetime distribution.

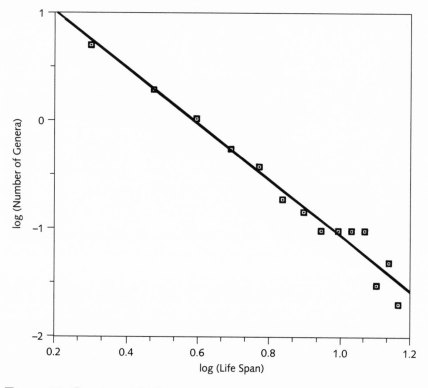

Figure 39. Continued (b) Log plot of the same data. The distribution is a power law with exponent roughly equal to 2.

Boettcher came up with a model in which each species is explicitly characterized by many traits, each of which gives a contribution to the fitness of the species. At each time step, the single trait with the lowest fitness among all the species is "mutated," that is, the corresponding fitness is replaced by a random number between 0 and 1. This trait interacts with one trait of the species to the right in a food-chain geometry and one trait to the left. Those traits are also assigned random new fitnesses. When there is exactly one trait for each species, the model reduces to the original punctuated equilibrium model.

Surprisingly, in the limit where there are many traits, the model can be solved exactly by very sophisticated mathematical methods. The distribution of avalanche sizes is a power law with exponent 3/2 just like Henrik's random neighbor model. The punctuated equilibrium evolution for a single species is

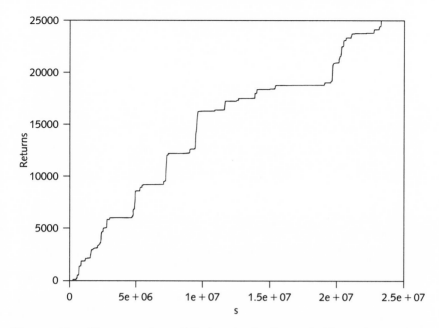

Figure 40.   Punctuated equilibrium in the Paczuski-Boettcher model. The curve shows the total number of mutation events for a single species vs. time. The distribution of the durations of the periods of stasis can be calculated rigorously. The exponent is 7/4.

depicted in Figure 40. The distribution of plateaus of the Devil's staircase is given by a power law with an exponent of 7/4.

The Self-Organization Process

The general process of self-organization in the punctuated equilibrium model has been studied by Maya Paczuski, Sergei Maslov, and myself. In contrast to sand models and earthquake models, it is possible to construct a mathematical theory for the slow process in which the ecology organizes itself to the critical state.

Sergei Maslov did his undergraduate studies at the prestigious Landau Institute in Moscow. His advisor was Valery Pokrovsky, famous for inventing the scaling theory of phase transitions on which essentially all our present understanding of critical phenomena is based, and for which, unexplicably, he did not receive the Nobel Prize. Our ideas of self-organized criticality (SOC) fall well in line with those ideas. During the last years of the Cold War, I had enjoyed visiting the Landau Institute many times, and developed some good friendships. Science, and physics in particular, enjoyed a good deal of respect and thrived quite well in the old Soviet Union. I had worked with Valery on a number of projects in condensed matter physics. Valery recommended Sergei to me, and we had him enroll at the State University of New York at Stony Brook, which is near Brookhaven National Laboratory. This led to his fruitful collaboration with Maya and me.

Ever since the inception of SOC, I had been frustrated by the lack of analytical (pen-and-paper) progress on SOC. Yes, indeed, there were the nice, exact results by Deepak Dhar, and beautiful approximate schemes for calculating the exponents, in particular by Luciano Pietronero's group in Rome, but there was essentially no progress on the important question of how the system becomes attracted to the critical state. However, this situation changed for the better in our collaboration with Sergei.

The approach to the critical point follows a characteristic pattern (Figure 30). The value of the largest fitness belonging to any species that has mutated up to a given time follows the stepwise curve shown in the figure. The steps of that curve show the points in time when that fitness grows. For a while after the step, there are lower fitnesses in the system, but eventually these low fitnesses are erased, and the curve has another small up-step. We call this curve the "gap" curve (and the equation that describes it the "gap" equation) since there are no species with fitnesses below the curve at the points in time when there is a step. The mutation activity between the steps are called "avalanches." The avalanches represent cascades of extinction events. One can show that the mutations during the avalanches are connected in a tree-like structure to the first mutation in the avalanche, that is, they are generated by a domino effect. After the completion of the avalanche where the curve makes a step, the activity jumps to somewhere else in the ecology, generally not connected with any species that mutated in the previous avalanche.

As the plateaus of the fitness curve reach higher and higher values, the avalanches, on average, become bigger and bigger. Eventually, the size of avalanches reaches infinity, limited only by the total number of species in the system, and the stepwise envelope curve ceases to increase. It gets stuck at the value $f_c = 0.667$. At that time the system has become critical and stationary. During the avalanches the fitnesses of some species are, by definition of the avalanches, less than the critical value, but at the end of an avalanche all fitnesses are again above the critical value. Thus, the self-organization can be described by an inescapable divergence of the size of avalanches. This divergence is described by a power law with an exponent gamma (γ), where $\gamma = 2.7$ in the model in which the interacting species were arranged on a circle.

The asymptotic approach of the gap f to the critical value as a function of time is yet another power law:

$$f(t) = f_c - A\left(\frac{t}{N}\right)^{-1/(\gamma-1)}.$$

Here, t is the total number of update steps, N is the number of species, and A is a constant factor. This equation is the fundamental equation for the process of self-organization. It shows that as t becomes larger and larger the gap f gets closer and closer to the critical value f_c. The envelope in Figure 30 follows that formula. The critical state with the unique value of the gap is an attractor for the dynamics, in contrast to non–self-organized critical systems where tuning is necessary. We call this equation the "gap equation."

A similar process is responsible for the criticality in sandpile models, although the insight here is mostly numerical. As the pile becomes steeper and steeper, the sand slides become larger and larger, until they reach the critical slope where they diverge and cover the entire system; this prevents further growth.

The Critical State

Once the system reaches the critical state, the evolutionary dynamics are described in terms of the spatiotemporal fractal shown in Figure 32. We have already defined the fractal dimension D of this fractal, and we have also defined the exponent τ for the avalanches. Interestingly, all other quantities that one might think of measuring can be expressed in terms of those two exponents. For instance, the exponent $\rho = 1/(\gamma - 1)$ in the gap equation for the relaxation of the critical state is a simple algebraic expression,

$\rho = (1 + 1/D - \tau)/(1 - 1/D)$. Another formula that we have derived allows us to determine the threshold very accurately. It turns out to be $f_c = 0.66700$, and not $2/3$ as we believed for a long time; it just happens to be very close.

Another quantity that we have ignored for some time is the power spectrum, i.e., the quantity that is supposed to show $1/f$-type noise. Again, we consider the mutation activity of a single species as time progresses. The punctuated equilibrium behavior, with periods of stasis of all durations separating bursts of activity, gives rise to a power spectrum $S(f) = 1/f^{\alpha}$, where the exponent $\alpha = 1 - 1/D$. For our model, the exponent is $0.58.$; for the Boettcher-Paczuski model the exponent can be found to be exactly $3/4$.

Thus, everything is quite well understood for the punctuated equilibrium model. The existence of the self-organized critical state has been proven. The resulting dynamics can be understood in terms of an underlying spatiotemporal fractal. The power spectrum is $1/f$-like; there are avalanches of all sizes. It provides insight into the origin of all the empirical results discussed in Chapter 1. Of course, our models are necessarily quite abstract, but they are robust. One can change features of the models without changing the criticality. This feature makes us believe that the models may be general enough to span the real world. As a fringe benefit, all the theoretical results hold for other models of self-organized criticality that are closely connected to phenomena such as fluid invasion and interface depinning.

Revisiting the Game of Life

Avalanches can be described by a simple terminology borrowed from particle physics. The species that have fitnesses below the threshold of 0.66700, shown in Figure 31, can be thought of as "particles." The avalanches can be thought of as cascade processes in which particles move, branch into more particles, or die. A particle moves when exactly one of the species to the right or to the left becomes a particle, that is, it gets a new fitness less than the critical value. A particle dies when all the species affected by the mutation process get fitnesses above the criticality. A particle branches into two or three particles if two or three of the species get fitnesses below criticality. An avalanche is over when there are no more particles. Then a new avalanche is initiated from a species with fitness at the critical value, and so on.

Recall how we studied the Game of Life. Starting from a static "dead" configuration, we started an avalanche of individuals coming and going, in a process that is entirely similar. Live sites may move, die, or branch out in the same way, until the Game of Life comes to rest in a new static state, and a new avalanche is initiated.

Particle physicists have constructed a theory for the phenomenon of cascade processes known as "Reggeon field theory" after its inventor, the particle physicist C. Regge of Italy. The theory describes a process in which particles can split up into other particles, and also annihilate each other. Reggeon field theory is not self-organized critical, but can be made critical by tuning the branching probability of the particles, just like a nuclear chain reaction. Maya had the idea that perhaps the critical behavior, and therefore the complexity of the Game of Life, can be understood from Reggeon field theory at its critical point, with the active sites having small fitnesses during the avalanches representing the particles.

We went to the library and found the best values for the avalanche distribution exponent in the two-dimensional version of Regge field theory, or "directed percolation," which it is also called. The value of the exponent was 1.28. To get the best possible value for the exponent in the Game of Life, we contacted two of our colleagues, Preben Alstrøm at the Niels Bohr Institute in Copenhagen, and Jan Hemmingsen at the German research facility in Juelich. They had made enormous numerical simulations similar to ours on the Game of Life, with avalanches extending to 100 million mutations. It is better to rely on someone else's results so as not to be prejudiced by our own ideas and wishes.

The results came back immediately: the value of the exponent indeed was 1.28! Thus, we had discovered a remarkable and very deep connection of Conway's Game of Life with its zoo of bizarre creatures, through our simulations of our evolution model, to the intricacies of particle physics. From the complex all the way to the simple.

Isn't this what science is all about? Relating hitherto disparate, seemingly unrelated phenomena to each other, thereby reducing the amount of unknown quantities in the world. We shall see yet another surprising example of this.

Revisiting Earthquakes

Very recently, things took yet another unexpected turn. Keisuke Ito of Kobe University in Japan, who was among the first to realize that earthquakes might be a self-organized critical phenomenon, made another interesting observation.

Ito realized that the punctuated equilibrium model can roughly be thought of as an earthquake model, simply by a change in terminology. The fitness landscape in the evolution model is equivalent to the heterogeneous barrier distribution over a fault plane that generates earthquakes. He had the two-dimensional version in mind, in which each species affects its four nearest neighbors. Mutation corresponds to rupture. In seismology, a nonuniform distribution of strengths over a fault plane is described in terms of "barriers" or "asperities," which are considered to cause the complex rupture process of earthquakes. The fitnesses in the evolution model can be thought of as the asperities in a fault model.

During an earthquake, a rupture starts from the weakest site in the crust with the minimum barrier strength. When the site breaks, the stress in the neighborhood changes. This can be modeled by assigning new random numbers to the new barriers at all those sites. Rupture propagates as long as the new barriers are weaker than the threshold for rupture. The earthquake stops when the minimum barrier becomes stronger than the threshold. Another earthquake starts from the site with the minimum barrier after some time when the tectonic stress is increased again. All these phenomena follow the punctuated equilibrium model.

To summarize, Ito views the entire dynamics of the fault zone as the dynamics of the evolution model depicted in Figure 32. We are dealing with one single dynamic process, not one process for each earthquake. Also the dynamics cannot be understood as a phenomenon associated with faults created by some independent process. The fault structure and the earthquakes are both generated by one process. There is only one spatiotemporal fractal structure. The spatial and temporal structures are two sides of the same coin. The temporal behavior at a specific site is given as a vertical cut in this fractal, and the spatial behavior is given as a horizontal cut.

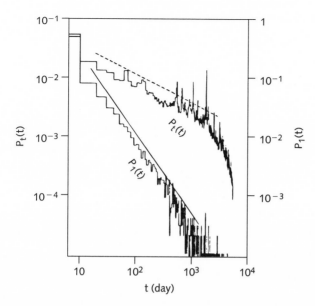

Figure 41. Distribution of waiting times ("first return" times), and "all return" times for earthquakes in California (1971–1985). The distribution of waiting times is a power law with exponent 1.4. This shows that earthquakes are a self-organized critical phenomenon (Ito, 1995).

How does this correspond to reality? Ito considered the time intervals it would take for earthquakes in California to return to the same small area, that is, he looked at the distribution of periods of stasis between earthquakes at a given location. He measured the distribution of these return times for 8,000 earthquakes. The result is shown in Figure 41. Strikingly, it is a power law with an exponent of 1.4, very similar to our exponent of 1.58. He also considered the distribution of times from a given earthquake to any subsequent earthquakes in the same region, not just the first earthquake. That is another power law, with exponent 0.5, compared with our exponent 0.42. Finally, he measured the distribution of spatial distances from one earthquake to the next consecutive one. That is another power law with exponent 1.7. The fact that there are power laws in both space and time suggests that there is one underlying space time fractal for the activity pattern of earthquakes in California, and that it is very possible that this fractal is generated by a dynamic process following rules similar to our evolution model.

The empirical result demonstrates that earthquakes are a self-organized critical phenomenon, with all of its hallmarks. The empirical power law for the return time, i.e., the periods of stasis, is interesting because it demonstrates that earthquakes are not periodic. There is a tendency, even among scientists, to view events that occur with some degree of regularity as periodic, as we have already seen in connection with Raup and Sepkoski's view of extinction data. The power law indicates that the longer you have waited since a large earthquake at a given location, the longer you can expect still to have to wait, contrary to common folklore. Earthquakes are clustered in time, not periodic.

The same goes for evolution. The longer a species has been in existence, the longer we can expect it to be around in the future. Cockroaches are likely to outlast humans.

I have often been asked what the realization that nature operates at a self-organized critical state is "good for." How can that help us predict or prevent earthquakes? How can I use it to make money on the stock market? If I am so smart, why am I not rich? Usually I don't like to answer these questions, not because I don't believe that the basic insight into how things work will not pay off at some time, but because I believe that acquiring insight is in itself a worthwhile effort.

There is one business that is entirely based on the statistical properties of events: the insurance business. I should be able to make a profit selling earthquake insurance! I would approach residents in earthquake areas where there has not been a major earthquake for a long time. The sales pitch would point out the "obvious" fact that an earthquake "is due"; nevertheless I would sell earthquake at a price that is lower than that of my competitors. On the other hand, I would stay away from areas where there has recently been a major earthquake.

the brain

The human brain is able to form images of the complex world surrounding us, so it might seem obvious that the brain itself has to be a complex object. However, it is not necessarily so. We have seen that complex behavior can arise from models with a simple architecture through a process of self-organization. Perhaps the brain is also a fairly simple organ.

Starting from a native state with little structure, the information about the surrounding environment is coded into the brain by a process in which the brain self-organizes into a critical state. In analogy with the sandpile, a "thought" may be viewed as a punctuation, i.e., a small or large avalanche triggered by some minor input in the form of an observation, or by another thought.

The brain contains trillions of neurons. Each neuron may be connected to thousands of other neurons. The firing mechanisms of individual neurons are fairly well understood, but how do trillions of neurons work together to form the emergent process we call thinking? Comparing with the way computers work, the function of the computer is not apparent from the properties of the individual transistors making up the computer. Those who construct computers do not even have to know how transistors work. The function of the computer comes from the way these interconnected transistors work together.

There is at least one major conceptual difference between the computer and the brain. The computer was built by design. An engineer put together all the circuits and made it work. No engineer—no computer. However, there is no engineer around to connect all the synapses of the brain. Even more to the point, there is no engineer available to make adjustments every time the outer world poses the brain with a new problem. One might imagine that the brain is ready and hard-wired from birth, with its connections formed through biological evolution, with all possible scenarios coded into the DNA. This does not make any sense. Evolution is efficient, but not that efficient. Indeed, the amount of information contained in the DNA is sufficient to determine general rules for neural connections but vastly insufficient to specify the whole neural circuitry. While there is some hard wiring—a lobster brain is different from a human brain—the functionality has to evolve during the lifetime of an individual. This means that the structure has to be self-organized rather than designed. Brain function is essentially created by the problems the brain has to solve.

Thus, to understand how the brain functions it is important to understand the process of self-organization. It is not enough to take the brain apart at some given instant and map out all existing connections, just as we don't understand the sandpile by just making a map of all the grains at some given point in time. Essentially all modeling of brain function from studying models of neural networks has ignored the self-organized aspects of the process, but has concentrated on designing a working brain model by engineering all the connections of inputs and outputs. This is good enough if the system is going to be used for some engineering purpose, such as pattern recognition, but it is basically misguided when it comes to understanding brain function.

Why Should the Brain Be Critical?

One may argue at least two different ways that the brain must be critical. First, consider a brain that is exposed to some external signal, representing for instance a visual image. The input signal must be able to access everything that is stored in the brain, so the system cannot be subcritical, in which case there would be access to only a small, limited part of the information. Grains dropped on a subcritical sandpile can only communicate locally by means of avalanches. The brain cannot be supercritical either, because then any input would cause an explosive branching process within the brain, and connect the input with essentially everything that is stored in the brain.

This can be seen in a different way. Consider a neuron somewhere in the brain, and an output neuron at a distance from that neuron. By changing the properties of the neuron, for instance by increasing or decreasing the strength of its connection with a neighbor neuron, it should be possible to affect the output neurons in the brain; otherwise that neuron would not have any meaningful function. If the brain is in the frozen subcritical state, there will be only a local effect of that change. If the brain is in a chaotic disordered state with neurons firing everywhere, it is not possible to communicate with the output neuron, and affect its signal properly, through all the noise.

Hence, the brain must operate at the critical state where the information is just barely able to propagate. At the critical state the system has a very high sensitivity to small shocks. A single grain of sand can lead to a very large avalanche. We say that a critical system has a large susceptibility. Of course, the avalanches at the critical state in the sandpile do not perform any meaningful function, so our problem is to teach the avalanches to connect inputs with the correct outputs.

How does the brain organize itself to the critical state? In the sand model, the criticality was ensured by adding grains of sand at a very slow rate, one grain at the time.

In the last couple of years I have been working on this problem together with Dimitris Stassinopoulos. Stassi had been working with Preben Alstrøm at the Bohr Institute on neural network models of steering processes, such as tracking a flying target. The network was kept at a critical state by a feedback mechanism that would keep the *output*, rather than the *input*, low.

It occurred to Stassi and me that maybe one could construct a toy brain model using ideas from that work, so I invited him to visit Brookhaven for a year.

The Monkey Problem

One of the problems in describing brain function is the uncertainty in determining what exactly is the problem that the brain is actually "solving." What, precisely is the function of the brain? It isn't enough to simply say that it is "thinking." A good deal of brain research traces the location of the activity of the brain when a person is subjected to various stimuli, but gives next to no insight on general principles. Before constructing a model, we found it important to define a specific problem that the brain was to solve.

A hungry monkey is confronted with the following problem. To get food, it must press one of two levers. At the same time it is shown a signal that can either be red or green. If the red signal is on, it has to press the left lever; if the green signal is on, it has to press the right lever. The signal switches back and forth between the red one and the green one. If the correct lever is pressed, the monkey will get a couple of peanuts.

A block diagram of the situation is shown in Plate 9. The outer world sends a signal to some of the neurons in the brain, through the eyes of the monkey. The resulting action of the monkey is fed back to the outer world, which in turn provides feedback to the monkey and its brain by either giving or denying food. After a number of wrong tries the monkey learns to perform properly.

The fact that the function of the brain has to be self-organized puts severe constraints on any brain model. In our model, neurons were arranged on a two-dimensional grid. The neurons in each row are connected with three neurons at the row below that neuron, as indicated by the arrows in the block diagram. We have also studied a network where the connections were completely random. This network functions almost as well, but is more difficult to illustrate graphically. The firing signal from the environment is represented by pulses that are fed into a number of random neurons, the red ones if the signal is red, the green ones if the signal is green. It is a simple matter to define the initial network. It took only a couple of sentences to specify the geometry. Not much more information is needed to specify a larger network. The brain model "at birth" is a simple structure.

At each step, each of the neurons are either in a "firing" state or a "nonfiring" state, according to whether their input voltage, or potential, exceeds a threshold. Firing neurons send electric signals to other neurons, driving their potentials closer to the threshold. This is very similar to the sand model, in which a toppling occurs if the height exceeds a threshold. The signal from a firing neuron is sent to the inputs of the three neurons in the layer below. The input of each neuron depends on the strengths of the connections between that neuron and the one that fired. In addition, a small amount of noise was added to all the inputs.

The output is given by the neurons in the bottom row. Say, for instance, that for the red signal neurons #10 and #15 counting from the left must fire, and for the green signal neurons #7 and #12 must fire (Plate 10).

In the beginning, the strengths of the connections between neurons were chosen arbitrarily. The red and green inputs were switched every 200 time steps, or whenever the output was correct. The feedback from the environment was sent to all the neuron connections in a totally democratic way. This could represent some hormone, or some other chemicals fed into the brain. In this sense, our model is fundamentally different from most other neural network models in which an amount of outside computation, not performed by the network itself, has to be carried out to update in detail the strength of the connections.

If there is a positive feedback signal, that is, the proper output neurons fired, all the connections between simultaneously firing neurons are strengthened whether or not they were responsible for the favorable result. If there is a negative signal, all these firing connections are weakened slightly. All other non-firing connections are left alone.

This type of scheme has been tried before without much success, precisely because of the weak communication between inputs and outputs, which makes learning prohibitively slow. Also, whenever the red signal is on, the pattern that is favorable for the green light is forgotten, and vice versa. An extra ingredient is needed. If there are too many output cells firing, all thresholds are raised. The function of this mechanism is to keep the activity as low as possible, and it results in setting up the brain in a critical state. To think clearly, you have to keep cool! If the activity becomes too low, for instance if the brain is asleep with no output, the thresholds of all neurons are lowered and more neurons fire. The monkey becomes hungry and activates the brain. Note that all of these processes are biologically reasonable; they can be performed by chemicals being sent around in the brain without a specific address.

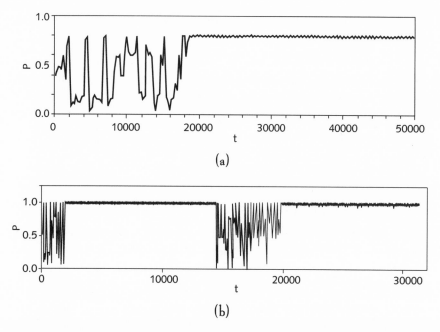

Figure 42. (a) Performance vs. the number of time steps. The performance is defined as the relative number of correct output cells that are firing. Note the transition where the brain has learned to switch quickly between the two patterns. In (b) a five by six block is removed after 15,000 time steps. After a while, the brain has learned again to switch correctly between the two outputs (Stassinopoulos and Bak, 1995).

Figure 42a shows the performance, measured as the relative number of output neurons that are firing correctly. After an initial period of very erratic response, which we call the learning period, the toy brain eventually learns to switch quickly back and forth between the correct responses. The transition is very sharp. Plate 10 shows the successful firing patterns as yellow squares for the two different inputs.

The Brain and River Networks

What happens inside the network during the learning process? Through a complicated organizational process, the system creates internal contacts or connections between selected parts of the input signal and the correct output cells. The process can be thought of as the formation of a river network with dams (thresholds). When the output is incorrect the riverbeds are raised (the firing connections are weakened) and the dams are lowered (the thresholds are lowered), which causes water to flow elsewhere; this is the process of "thinking." During that period there is an increased activity in the brain.

If at some point the response happens to include the correct cells, the riverbeds are deepened but all the dams are raised, which prevents the signal from going elsewhere. The system tries to lower the activity as much as possible while still connecting with the correct output. At some point the thresholds will become too high and the output becomes too low (the monkey loses concentration), but the system immediately responds by lowering the threshold. The small amount of random noise prevents the network from locking into wrong patterns, with too deep riverbeds, from which it cannot escape. It allows the network to explore new possibilities. Each input fills up a river network shown by the yellow squares.

The process is somewhat similar to the mechanism for the evolution model. During periods of low fitness (improper connections with output) there is a relatively great activity where the system tries many different connections until it finds a state with correct connections (high fitness), in which most of the neurons are passive, just as the species in the evolution model have fitnesses above threshold in the periods of stasis, in which they have "learned" to connect properly with the environment.

The ability of fast switching is related to the system operating at the critical point. The signal is barely able to propagate through the system. The flow pattern is very similar to Andrea Rinaldo's critical river networks, and does not look like a flooded system with large lakes. A traditional neural network model corresponds to a flooded system with roughly half the neurons firing at all times, resulting in poor communication. At the critical point, the system can easily switch from a state in which one system of rivers is flowing to a state in which another system of rivers is flowing to a different output. We exploit the high susceptibility of the critical state.

The network is robust to damage. After 150,000 steps, a block of thirty neurons was removed from the network as shown in Plates 10c and 10d. After a transient period the network had relearned the correct response by carving new rivers in the network. The performance is shown in Figure 42b. In other words, instead of using some features of the input signal the system learns to use other features. Think of this as replacing "vision" with "smell."

Our toy brain model is not a realistic model of the brain. Its sole purpose is to demonstrate that aspects of brain function can be understood from a system starting with a minimum amount of structure. The ability of the brain to function is intimately related to its dynamics, which organize knowledge about the outer world into critical pathways of firing networks in an otherwise quiet medium. The criticality allows for fast switching between different complicated patterns without interference. The memory is encoded as a network of riverbeds waiting to be filled up under the relevant external stimulus.

The feedback between reality and the individual through perception of the physical world determines the interconnected structure of the brain.

on economics
and traffic jams

So far we have proceeded from astrophysics to geophysics, and from geophysics to biology and the brain. We now take yet another step in the hierarchy of complete phenomena, into the boundary between the natural world and the social sciences. Humans interact with one another. Is it possible that the dynamics of human societies are self-organized critical? After all, human behavior is a branch of biology, so why should different laws and mechanisms be introduced at this point? Here two specific human activities will be considered, namely economics and traffic. Perhaps these phenomena are simpler than other human activities. At least, the activities can be quantified and measured, in terms of prices, volumes, and velocities. That might be the reason that economics exists as a discipline independent of other social sciences.

Equilibrium Economics
Is Like Water

Traditional equilibrium interpretations of economics resemble the description of water flowing between reservoirs. Goods and services flow easily from agent to agent in amounts such that no further flow or trade can be advantageous to any trading partner. A small change in the economy, such as a change in the interest rate, causes small flows that adjust the imbalance.

Specifically, consider two agents each having a number of apples and oranges. One has many oranges and few apples, the other has many apples and few oranges. Since having too many apples or oranges may not be desirable, they trade some of their apples and oranges. Before trading, oranges are worth more for the agent with too few oranges than for the other agent. They trade a precise amount such that oranges are worth exactly the same number of apples for the two agents, which removes any incentive for further trading. At that point it is not to anybody's advantage to trade further. The agents are perfectly rational, so they both know how many apples to buy and sell, and what the exchange rate should be. They are perfectly predictable. In our water flow analogy, the water in two connected glasses of water will flow from one glass to the other until the equilibrium point where the levels in the two glasses are the same.

In equilibrium systems, everything adds up nicely and linearly. It is trivial to generalize to many agents; this simply corresponds to connecting more glasses of water. The effect on the water level from adding several drops of water is proportional to the number of drops. One does not have to think about the individual drops. In physics, we refer to this kind of treatment where only a global macrovariable, such as the water level, is considered as a "mean field approximation." Traditional economics theories are mean field theories in that they deal with macrovariables, such as the the gross national product (GNP), the unemployment rate, and the interest rate. Economists develop mathematical equations that are supposed to connect these variables. The differences in individual behavior average out in this kind of treatment. No historical accident can change the equilibrium state, since the behavior of rational agents is unique and completely defined. Mean field treatments work quite well in physics for systems that are either very ordered or very disordered. However, they completely fail for systems that are at or near a critical state. Unfortunately, there are many indications that economics systems are in fact critical.

Traditional economics does not describe much of what is actually going on in the real world. There are no stock market crashes, nor are there large fluctuations from day to day. Contingency plays no role in perfectly rational systems in which everything is predictable.

Equilibrium economics does not even work for the simple example of the agents trading apples and oranges. Neither one knows how much oranges and apples are worth for the other agent. When offering apples for sale, they may sell too cheaply, or ask too high a price, so that the proper equilibrium will never be reached. They may end up with more apples than they want. Agents in reality are not perfectly rational. In discussions with traditional economists, I used to argue that their economics theory has to include me, and that I certainly am not perfectly rational, as they themselves have argued so convincingly.

The obsession with the simple equilibrium picture probably stems from the fact that economists long ago believed that their field had to be as "scientific" as physics, meaning that everything had to be predictable. What irony! In physics detailed predictability has long ago been devalued and abandoned as a largely irrelevant concept. Economists were imitating a science whose nature they did not understand.

Perfect rationality makes things nice and predictable. Without this concept, how can you characterize the degree of ignorance among agents, and how can you then predict anything? I first faced this stubbornness to give up the ideas of perfect rationality during my first visit to Santa Fe. During lunch at the Coyote Cafe with a variety of scientists visiting the institute, including one of the foremost and smartest classic economists, Michele Boldrin, I discussed the absurdity of the "perfect rationality concept" in a world of real people. All the time Boldrin was nodding and saying yes, yes, yes, to all the arguments. The discussion continued as we were walking back to the institute. However, just as we were turning into the courtyard Michele concluded, "I still prefer the 'perfect rationality' concept."

Real Economics Is Like Sand

But economics is like sand, not like water. Decisions are discrete, like the grains of sand, not continuous, like the level of water. There is friction in real economics, just like in sand. We don't bother to advertise and take our apples to the market when the expected payoff of exchanging a few apples and oranges is too small. We sell and buy stocks only when some threshold price is reached, and remain passive in between, just as the crust of the earth is stable until the force on some asperity exceeds a threshold. We don't continually adjust our holdings to fluctuations in the market. In computer trading, this threshold dynamics has been explicitly programmed into our decision pattern. Our decisions are sticky. This friction prevents equilibrium from being reached, just like the friction of sand prevents the pile from collapsing to the flat state. This totally changes the nature and magnitude of fluctuations in economics.

Economists close their eyes and throw up their hands when it comes to discussing market fluctuations, since there cannot be any large fluctuations in equilibrium theory. "Explanations for why the stock market goes up or down belong on the funny pages," says Claudia Goldin, an economist at Harvard. If this is so, one might wonder, what do economists explain?

The various economic agents follow their own, seemingly random, idiosyncratic behavior. Despite this randomness, simple statistical patterns do exist in the behavior of markets and prices. Already in the 1960s, a few years before his observations of fractal patterns in nature, Benoit Mandelbrot analyzed data for fluctuations of the prices of cotton and steel stocks and other commodities. Mandelbrot plotted a histogram of the monthly variation of cotton prices. He counted how many months the variation would be 0.1% (or −0.1%), how many months it would be 1%, how many months it would be 10%, etc. (Figure 3). He found a "Levy distribution" for the price fluctuations. The important feature of the Levy distribution is that it has a power law tail for large events, just like the Gutenberg–Richter law for earthquakes. His findings have been largely ignored by economists, probably because they don't have the faintest clue as to what is going on.

Traditionally, economists would disregard the large fluctuations, treating them as "atypical" and thus not belonging in a general theory of economics. Each event received its own historical account and was then removed from the data set. One crash would be assigned to the introduction of program trading, another to excessive borrowing of money to buy stocks. Also, they would "detrend" or "cull" the data, removing some long-term increase or decrease in the market. Eventually, they would end up with a sample showing only small fluctuations, but also totally devoid of interest. The large fluctuations were surgically removed from the sample, which amounts to throwing the baby out with the bathwater. However, the fact that the large events follow the same behavior as the small events indicates that one common mechanism works for all scales—just as for earthquakes and biology.

How should a generic model of an economy look? Maybe very much like the punctuated equilibrium model for biological evolution described in Chapter 8. A number of agents (consumers, producers, governments, thieves, and economists, among others) interact with each other. Each agent has a limited set of options available. He exploits his options in an attempt to in- crease his happiness (or "utility function" as the economists call it to sound more scientific), just as biological species improve their fitness by mutating. This affects other agents in the economy, who now adjust their behavior to the new situation. The weakest agents in the economy are weeded out and re- placed with other agents, or they modify their strategy, for instance by copy- ing more successful agents.

This general picture has not been developed yet. However, we have con- structed a simplified toy model that offers a glimpse of how a truly interactive, holistic theory of economics might work.

Simple Toy Model of a Critical Economy

A couple of days after my introductory talk at the Santa Fe Institute in 1988, Michael Woodford and Jose Scheinkman of the University of Chicago en- tered my office at the institute and wanted to discuss a sandpile-type model of economics. Mike is an economist belonging to the traditional school, very clever and very conservative, whereas Jose had already attempted to apply ideas from chaos theory to economics. They sketched their ideas on the black- board, and I became very enthusiastic.

Their idea was to construct a simplified network of consumers and pro- ducers. This led to a very productive, though rather painful, collaboration, reflecting the very different modes of operating in physics and economics.

Theoretical economists like to deal only with models that can be solved analytically with pen and paper mathematics. I always found this quite ironic. Physics is a much simpler science than economics, but nevertheless very rarely are we able to "solve" problems in the mathematical sense. Even the world's most sophisticated mathematics is insufficient to deal rigorously with many problems in physics. Sometimes we use numerical simulations; sometimes we use approximate theories. Surely, some of these approximations must look horrifying to a pure mathematician. However, although sometimes based on sheer intuition, they work well and provide a good deal of insight into the relevant physics. The physicist performs one dirty mathematical trick after another. Invariably, however, there is a mathematician running after him, who will eventually almost catch up with him and yell, "It was all right what you did!"

It appears to me that economics, because of the complexity of the systems involved, does not call for exact mathematical solutions. Indeed, the model that we came up with, despite its simplicity, could not be solved mathematically. I went back to Brookhaven, where Kan Chen, the research associate with whom I did the simulation of the Game of Life, performed numerical simulations on the model. Indeed, the model was critical, with avalanches of all sizes. However, Mike was quite uncomfortable with the numerical nature of the solution, and Kan Chen and I continued working on the problem until we, alas, did come up with a model that could be solved mathematically without sacrificing the scientific content, to everybody's satisfaction.

The model is illustrated in Figure 43. It is a network of producers, who each buy goods from two vendors, produce goods of their own, and sell them to two customers. The producers may start the process by having random amounts of goods stored, or they may start with nothing. It makes no difference. At the beginning of each period, say each week, the producers receive orders of one or zero units from each customer. If they have a sufficient amount of goods in stock, they transfer it to the customers; if not, they send orders to their two vendors, receive one unit from each, and produce two units to fill the order. If they have one unit left after these transactions, they store it until next week. Each producer thus plays the dual role of selling to his customers, and buying from his vendors. The process starts at the upper row in the network,

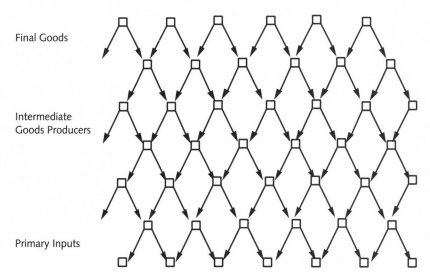

Figure 43.   Model of interacting producers. Each producer receives orders from two customers in the row above him. If he does not have sufficient amounts of goods in stock, he sends orders to two vendors further down the network, receives one unit of goods from each vendor, produces two units of his own goods, ships the ordered amount of goods. If he has one unit of his goods left after the transactions, he keeps it in stock for the next round. The process is initiated from consumer demands on top of the diagram (Bak et al., 1993).

which represent consumers, and ends at the bottom row, which represents the producers of raw materials.

First, we considered the situation where each week there would be a single "shock" triggering the economy, with only one consumer demanding goods. This initial demand leads to a "trickle-down" effect in the network. Figure 44 shows a typical state of the network, with each producer marked by the number of goods he has in stock after completing the previous week's trades. An empty circle indicates zero units, a full circle one unit. The first vendor has nothing in stock. He receives two units from his vendors, sells one unit to the consumer, and keeps one unit in stock for the next week. His vendors actually did not have the demanded products in stock, and had to send orders further down in the network. After a number of events, the avalanche stops. The figure shows the extent of the avalanche, and the amount the vendors have in

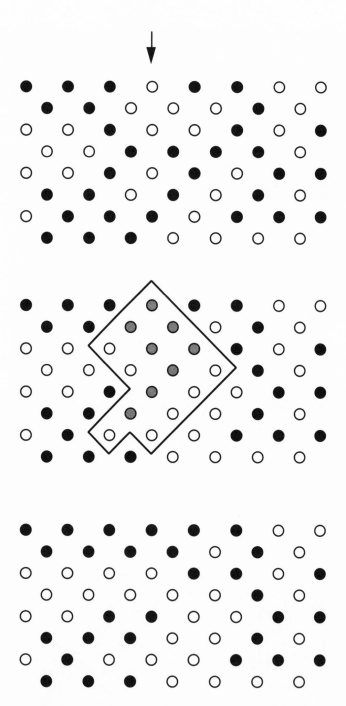

Figure 44. Network (a) before and (c) after an avalanche initiated by a single demand at the position of the arrow. The arrow indicates the flow of orders. The goods flow in the opposite direction. The black dots indicate the agents who have one unit of their goods in stock. The gray dots indicate agents who had to produce in order to fulfill the demands. The enclosed area indicates the size of the avalanche (Bak et al., 1993).

stock at the end of the week. Thus, small shocks can lead to large avalanches. The contribution of the event to the GNP is the area of the avalanche, that is the total amount of goods produced during the avalanche.

We could solve the model because we could relate it to another model that had previously been solved by Deepak Dhar and Ramakrishna Ramaswamy at the Tata Institute in Bombay, in the context of sandpiles. The model is directional, in the sense that information is only transmitted down in the network, not up. Dhar and Ramaswamy showed that the distribution of avalanches is a power law, $N(s) = s^{-\tau}$, with $\tau = \frac{4}{3}$.

To go from the power law to the Levy law observed by Mandelbrot, all that one has to do is to consider the situation in which each week there are several customers, and not just one, each demanding final goods. Each demand leads to an avalanche, so each day there are many avalanches of different sizes. One can show rigorously that for very many customers the result of the distribution of the total activity is the Levy function. I was able to demonstrate this by means of a simple mathematical calculation that would satisfy any physicist's demand for rigor. Nevertheless, my methods did not satisfy my very demanding collaborators, who didn't yield before they found the formula for how to add power law distributed random variables to arrive at the Levy distribution in a mathematics textbook.

Fluctuations and Catastrophes Are Unavoidable

Our conclusion is that the large fluctuations observed in economics indicate an economy operating at the self-organized critical state, in which minor shocks can lead to avalanches of all sizes, just like earthquakes. The fluctuations are unavoidable. There is no way that one can stabilize the economy and get rid of the fluctuations through regulations of interest rates or other measures. Eventually something different and quite unexpected will upset any carefully architectured balance, and there will be a major avalanche somewhere else in the system.

In contrast to our critical economy, an equilibrium economy driven by many independent minor shocks would show much smaller fluctuations.

Those fluctuations are given by a Gaussian curve, better known as the "bell curve" which has negligible tails. There is no possibility of having large fluctuation or catastrophes in an equilibrium economy.

Although economists do not understand the large fluctuations in economics, the fluctuations are certainly there. Karl Marx saw these fluctuations in employment, prices, and production as a symbol of the defunct capitalistic society. To him, the capitalistic society goes from crisis to crisis. A centralized economy would eliminate the fluctuations to the benefit of everybody, or at least the working class. Marx argued that a large avalanche, namely a revolution, is the only way of achieving qualitative changes.

Alan Greenspan, chairman of the Federal Reserve, manipulates the interest rate in order to avoid inflationary bursts—even with the prospect of slowing down the economy. Common to Greenspan's and Marx's view is the notion that fluctuations are bad and should be avoided in a healthy economy.

If economics is indeed organizing itself to a critical state, it is not even in principle possible to suppress fluctuations. Of course, if absolutely everything is decided centrally, fluctuations could be suppressed. In the sandpile model, one can carefully build the sandpile to the point where all the heights are at their maximum value, $Z = 3$. However, the amount of computations and decisions that have to be done would be astronomical and impossible to implement. And, more important, if one indeed succeeded in building this maximally steep pile, then any tiny impact anywhere would cause an enormous collapse. The Soviet empire eventually collapsed in a mega-avalanche (not predicted by Marx). But maybe, as we shall argue next in a different context, the most efficient state of the economy is one with fluctuations of all sizes.

Traffic Jams

Taking a broader view, economics deals with the way humans interact, by exchanging goods and services. In the real world, each agent has limited choices, and a limited capability of processing the information available; he has "bounded rationality." In some sense, the situation of the individual agent is like a car driver in traffic on a congested highway. His maximum speed is limited by the cars in front of him (and perhaps by the police); his distance to the car in front of him is limited by his ability to brake. He is exposed to random shocks from the mechanical properties of his car and from bumps in the road.

Kai Nagel and Michael Schreckenberg of the University of Duisburg, Germany have constructed a simple cellular automaton model for one-lane highway traffic along those lines. Cars can move with velocity 0, 1, 2, 3, 4, or 5. This velocity defines how many "car lengths" each car will move at the next time step. If a car is moving too fast, it must slow down to avoid a crash. A car that has been slowed down by a car in front will accelerate again when given an opportunity. The ability to accelerate is less than the ability to break, that is it takes more time steps to go from 0 to 5, than to brake from 5 to 0. Depending on the total number of cars on the road, there are two possible situations. If there are few cars there is a free flow of cars with only small traffic jams. If the density is high there is massive congestion.

Kai Nagel came to visit us a couple of years ago, while still a graduate student in Germany. Kai had already carried out a theoretical study in meteorology, arguing that the formation of fractal clouds is a self-organized critical process. Maya Paczuski and Kai considered the traffic emerging from one large traffic jam. Think of the Long Island Expressway, which runs along Long Island, starting at the Queens Midtown Tunnel leading into Manhattan. They came up with a theory that could describe the traffic coming out of the tunnel in the rush hour, where the largest possible number of cars would be pumped into the expressway. Everybody living on Long Island is familiar with the resulting huge traffic jams that can occur on the expressway, which has been called "the world's largest parking lot."

Figure 45 shows the computer-simulated traffic jams. The horizontal axis is the highway, the vertical axis is time. Time is increasing in the downward direction. The cars are shown as black dots. The cars originate from a huge jam to the left, which is not seen, and all move to the right. The diagram allows us to follow the pattern in time and space of the traffic. At each time step, the position of each car is shifted to the right by an amount equal to the velocity of that car. Traffic jams are shown as dense dark areas, where the distance between the cars is small. Also, the positions of cars between two successive horizontal lines are only shifted slightly since their velocities are low.

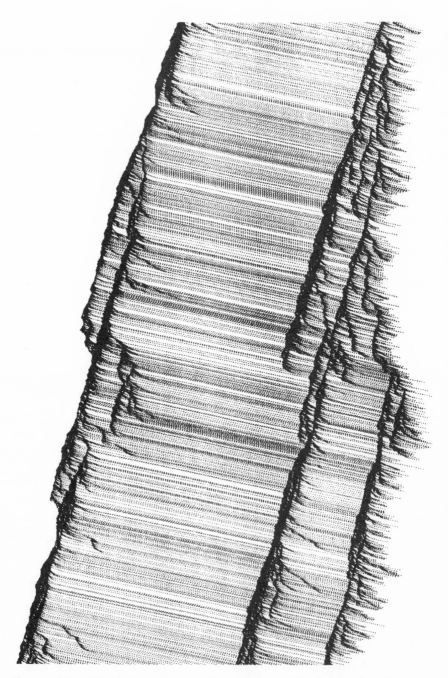

Figure 45.   Traffic jam simulated by computer. The horizontal direction indicates a highway. Cars are shown as black dots. Time progresses in the downward direction. The dots form trajectories of the individual cars, which appear as lines. The dark areas with a high density of cars indicate traffic jams. The pattern was set up from a huge jam at the left pumping cars into the highway at the maximum rate. The emergent jam on the right was initiated by slowing down one car in the top right-hand corner (Nagel and Paczuski, 1995).

Traffic jams may emerge for no reason whatsoever! They are "phantom" traffic jams. A small random velocity reduction from 5 to 4 of a single car is enough to initiate huge jams. We have met the same situation before: for earthquakes, for biological evolution, for river formation, and for stock market crashes. A cataclysmic triggering event (like a traffic accident) is not needed. Our natural intuition that large events come from large shocks has been violated. It does not make any sense to look for specific reasons for the jams.

The jams are fractal, with small subjams inside big jams ad infinitum. This represents the irritating stop-and-go driving pattern that we are all familiar with in congested traffic. On the diagram, it is possible to trace the individual cars and observe the stop-and-go behavior.

Traffic jams move backward, not forward, as can be seen in the figure. For comparison, a similar diagram for the traffic on a real highway in Germany is also showed in Figure 46. The picture is based on photos of the highway taken at regular intervals. Note that the general features are the same as for the computer simulations, including the backward motion of the jams. Eventually, the jams dissolve. From extensive computer simulations, the number of traffic jams of each size was calculated. Of course (you guessed it) they found a power-law distribution. The exponent for the power law appeared to be close to $3/2$. This suggested an elegant but simple theory of the phenomenon, a "random walk theory."

Each jam starts at a random nucleation point, at the top of the figure. At each time step, the size of the jam can either increase, with a certain probability, or decrease, with the same probability. Because of this 50-50 situation, the process is critical. This process can be solved mathematically, and gives a power law distribution, with an exponent that is exactly 1.5 as suggested by the simulations. We have met the random walk picture of self-organized critical systems before, in the context of the random neighbor model of evolution in Chapter 9.

Highway traffic is a classic example of $1/f$ noise. Over 20 years ago, T. Musha and H. Higuchi measured the flow of traffic on the Kanai Expressway in Japan as a function of time, by standing on a bridge over the highway and measuring the times that cars passed under the bridge. They observed a curve similar to that of light from a quasar. When measuring the power spectrum,

Space (Road)

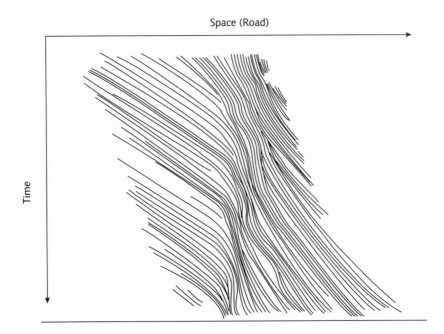

Time

Figure 46.   Traffic jam on a German highway from aereal photography. The plot is similar to the one from the numerical simulation in Figure 45, with each line representing the motion of one vehicle (Treiterer).

they found components of all frequencies, with the famous $1/f$ distribution. Kai and Maya did the same measurement on the computer-simulated traffic data. Standing on a bridge and monitoring the traffic corresponds to measuring the patterns of black dots along a vertical line. The signal is a Devil's staircase, just as for the evolution model. They also found a $1/f^{\alpha}$ noise (Figure 47) in the computer simulations. Moreover, they were able to prove mathematically that $\alpha = 1$ from a cascade mechanism, where subjams at each time step can grow or die or branch into more jams. For once, we have an accurate and complete understanding of the elusive $1/f$ noise in a model system that actually describes reality. As for the other phenomena that we have studied, $1/f$ noise is due to scale-free avalanches in a self-organized critical system. In the case of traffic, the $1/f$ noise is the mathematical description of the irritating, unpredictable stop-and-go behavior in traffic jams.

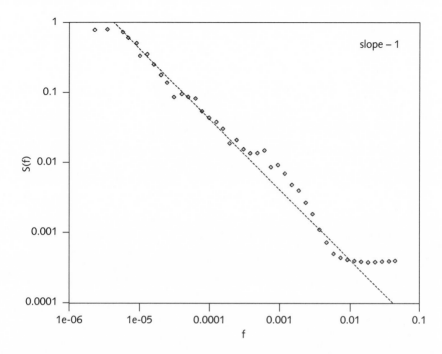

Figure 47. Power spectrum for traffic jam (Nagel and Paczuski, 1995).

Kai and Maya studied the situation in which there were only very rare random fluctuations initiating traffic jams. Interestingly, they point out that technological advancements such as cruise control or radar-based driving support would tend to reduce the fluctuations around maximum speed, and thus increase the range of validity of their results. One unintended consequence of these flow control technologies is that, if they work, they would in fact push the traffic system closer to its underlying critical point, thereby making prediction, planning, and control more difficult, in sharp contrast to the original intentions. Note the analogy with attempts to regulate economy (or sandpiles). Self-organized criticality is a law of nature for which there is no dispensation.

They made one final observation. Traffic jams are a nuisance, amplified by our lack of ability to predict them. Sometimes we are slowed down by a large jam, sometimes we are not. One might suspect that there would be a more efficient way of dealing with the traffic. In fact there might not be.

The critical state, with jams of all sizes, is the most efficient state. The system has self-organized to the critical state with the highest throughput of cars. If the density were slightly less, the highway would be underutilized, if the density were slightly higher, there would be one big permanent, huge jam, absorbing a fraction of the cars. In both cases, the throughput would be less.

More precisely, the critical state is the most efficient state *that can actually be reached dynamically.* A carefully engineered state where all the cars were moving at velocity 5 would have higher throughput, but it would be catastrophically unstable. This very efficient state would collapse long before all the cars became organized.

This gives some food for thought when applied to economics in general. Maybe Greenspan and Marx are wrong. The most robust state for an economy could be the decentralized self-organized critical state of capitalistic economics, with fluctuations of all sizes and durations. The fluctuations of prices and economic activity may be a nuisance (in particular if it hits you), but that is the best we can do!

The self-organized critical state with all its fluctuations is not the best possible state, but it is the best state that is dynamically achievable.

Bibliography

Chapter 1

Benton, M. J. ed. *The Fossil Record 2*. London: Chapman and Hall, 1993.

Dawkins, R. *The Blind Watchmaker*. New York: Penguin, 1988.

Feder, J. *Fractals*. New York: Plenum, 1988.

Feigenbaum, M. J. Quantitative Universality for a Class of Nonlinear Transformations. *Journal of Statistical Physics* 19 (1978) 25.

Gleick, J. *Chaos*. New York: Viking, 1987.

Gould, Stephen Jay. *Wonderful Life*. New York: Norton, 1989.

—— and Eldridge, N. Punctuated Equilibrium: The Tempo and Mode of Evolution Reconsidered. *Paleobiology* 3 (1977) 114.

Gutenberg, B. and Richter, C. F. *Seismicity of the Earth*. Princeton, NJ: Princeton University Press, 1949.

Johnston, A. C. and Nava, S. Recurrence Rates and Probability Estimates for the New Madrid Seismic Zone. *Journal of Geophysical Research* 90 (1985) 6737.

Mandelbrot, B. The Variation of Certain Speculative Prices. *Journal of Business of the University of Chicago* 36 (1963) 307.

——. The Variation of Some Other Speculative Prices. *Journal of Business of the University of Chicago* 37 (1964) 393.

——. How Long is the Coast of Britain? *Science* 156 (1967) 637.

——. *The Fractal Geometry of Nature.* New York: Freeman, 1983.

Officer, C. and Page, J. *Tales of the Earth. Paroxysms and Perturbations of the Blue Planet.* Oxford, New York: Oxford University Press, 1993.

Press, W. H. Flicker Noise in Astronomy and Elsewhere. *Comments on Astrophysics* 7 (1978) 103.

Prigogine, I. *From Being to Becoming.* San Francisco: Freeman, 1980.

Raup, D. M. Biological Extinction in Earth History. *Science* 231 (1986) 1528.

——. *Extinction: Bad Genes or Bad Luck.* New York: Norton, 1991.

Raup, D. M. and Sepkoski, J. J. Periodicity of Extinctions in the Geological Past. *Proceedings of the National Academy of Science, USA* 81 (1984) 801.

Ruderman, D. L. The Statistics of Natural Images. *Network: Computation in Neural Systems* 5 (1994) 517.

Schroeder, M. *Fractals, Chaos, Power Laws.* New York: Freeman, 1991.

Sepkoski, J. J. Jr. Ten Years in the Library: New Data Confirm Paleontological Patterns. *Paleobiology* 19 (1993) 43.

——. *Mass Extinction Processes: Periodicity.* In: Briggs, D. E. G. and Crowther, P. R., eds. *Paleobiology.* Oxford: Blackwell, 1992:171.

Zipf, George Kingsley. *Human Behavior and the Principle of Least Effort.* Cambridge MA: Addison-Wesley, 1949.

Chapter 2

Bak, P., Tang, C. and Wiesenfeld, K. Self-Organized Criticality. An Explanation of $1/f$ Noise. *Physical Review Letters* 59 (1987) 381.

——. Self-Organized Criticality. *Physical Review A* 38 (1988) 364.

Chapter 3

Bak, P. and Chen, K. Self-Organized Criticality. *Scientific American* 264, Janurary (1991) 46.

Bak, P. and Creutz, M. Fractals and Self-Organized Criticality. In: *Fractals and Disordered Systems* 2. Bunde, A. and Havlin, S., eds. Berlin, Heidelberg: Springer, 1993.

Bak, P. and Paczuski, M. Why Nature Is Complex. *Physics World* 6 (1993) 39.

——. Complexity, Contingency, and Criticality. *Proceedings of the National Academy of Science, USA* 92 (1995) 6689.

Chhabra, A. B., Feigenbaum, M. J., Kadanoff, L. P., Kolan, A. J., and Procaccia, I. Sandpiles, Avalanches, and the Statistical Mechanics of Non-equilibrium Stationary States. *Physical Review E* 47 (1993) 3099.

Christensen, K., Fogedby, H. C., and Jensen, H. J. Dynamical and Spatial Aspects of Sandpile Cellular Automata. *Journal of Statistical Physics* 63 (1991) 653.

Dhar, D. Self-Organized Critical State of Sandpile Automata Models. *Physical Review Letters* 64 (1990) 1613.

Gore, A. *Earth in the Balance.* Boston: Houghton Mifflin, 1992.

Chapter 4

Bretz, M., Cunningham, J. B., Kurczynsky, P. L. and Nori, F. Imaging of Avalanches in Granular Materials. *Physical Review Letters* 69 (1992) 2431.

Frette, Vidar, Christensen, Kim, Malthe-Sørenssen, Anders, Feder, Jens, Jøssang, Torstein, and Meakin, Paul. Avalanche Dynamics in a Pile of Rice. *Nature* 379 (1995) 49.

Held, G. A., Solina, D. H., Keane, D. T., Haag, W. J., Horn, P. M., and Grinstein, G. Experimental Study of Critical Mass Fluctuations in an Evolving Sandpile. *Physical Review Letters* 65 (1990) 1120.

Jaeger, H. M., Liu, C., and Nagel S. R. Relaxation of the Angle of Repos. *Physical Review Letters* 62 (1989) 40.

Jaeger, H. M. and Nagel, S. R. Physics of the Grandular State. *Science* 255 (1992) 1523.

Noever, D. A. Himalayan Sandpiles. *Physical Review E* 47 (1993) 724.

Rigon, R., Rinaldo, A., and Rodriguez-Iturbe, I. On Landscape Selforganization. *Journal of Geophysical Research* 99 (1994) 11971.

Rinaldo, A., Maritan, A., Colaiori, F., Flammini, A., Rigon, R., Ignacio, I., Rodriguez-Iturbe, I., and Banavan, J. R. Thermodynamics of Fractal River Networks, *Physical Review Letters* 76, 3364 (1996).

Rothman, Daniel H., Grotzinger, John P., and Flemings, Peter. Scaling in Turbidite Deposition. *Journal of Sedimentary Research A* 64 (1994) 59.

Somfai, E., Czirok, A., and Vicsek, T. Self-Affine Roughening in a Model Experiment on Erosion in Geomorphology. *Journal of Physics A* 205 (1994) 355.

———. Power-Law Distribution of Landslides in an Experiment on the Erosion of a Granular Pile. *Journal of Physics A* 27 (1994) L757.

Turcotte, D. L. *Fractals and Chaos in Geology and Geophysics.* Cambridge, England: Cambridge University Press, 1992.

Yam, Philip. Branching Out. *Scientific American* 271, November (1994) 26.

Chapter 5

Bak, P. and Chen, K. *Fractal Dynamics of Earthquakes* In: Barton, C. C. and Lapointe, P. R., eds. *Fractals in the Earth Sciences.* New York: Plenum, 1994.

Bak, P. and Tang, C. Earthquakes as a Self-Organized Critical Phenomenon. *Journal of Geophysical Research* B94 (1989) 15635.

Burridge, R. and Knopoff, L. *Bulletin of the Seismological Society of America* 57 (1967) 341.

Carlson, J. M. and Langer, J. S. Physics of Earthquakes Generated by Fault Dynamics. *Physical Review Letters* 62 (1989) 2632; *Physical Review A* 40 (1989) 6470.

Chen, K., Bak, P., and Obukhov, S. P. Self-Organized Criticality in a Crack Propagation Model of Earthquakes. *Physical Review A* 43 (1990) 625.

Dennis, B. R. *Solar Physics* 100 (1985) 65.

Diodati, P., Marchesoni, F., and Piazza, S. Acoustic Emission from Volcanic Rocks: An Example of Self-organized Criticality. *Physical Review Letters* 67 (1991) 2239.

Garcia-Pelayo, R. and Morley, P. D. Scaling Law for Pulsar Glitches. *Europhys. Lett.* 23 (1993) 185.

Ito, K. and Matsuzaki, M. Earthquakes as Self-Organized Critical Phenomena. *Journal of Geophysical Research* B 95 (1990) 6853.

Lu, E. T. and Hamilton, R. J. Avalanches and the Distribution of Solar Flares. *Astrophysical Journal* 380 (1991) L89.

Lu, E. T., Hamilton, R. J., McTiernan, J. M., and Bromund, K. R. Solar Flares and Avalanches in Driven Dissipative Systems. *Astrophysical Journal* 412 (1993) 841.

Luongo, G., Mazzarella, A., and Palumbo, A. On the Self-Organized Critical State of Vesuvio Volcano. *Journal of Volcanology and Geothermal Research* 70, (1996) 67.

Mineshige, S., Takeuchi, M., and Nishimori, H. Is a Black-Hole Accretion Disc in a Self-Organized Critical State? *Astrophysical Journal* 435 (1994) L125.

Morley, P. D. and Smith, I. Platelet Model of Pulsar Glitches. *Europhysics Letters* 33 (1996) 105.

Olami, Z., Feder, H. J., and Christensen, K. Self-Organized Criticality in a Continuous, Nonconservative Cellular Automaton Modeling Earthquakes. *Physical Review Letters* 68 (1992) 1244.

Sornette, A. and Sornette, D. Self-Organized Criticality and Earthquakes. *Europhysics Letters* 9 (1989) 197.

Vieira, M. de Sousa. Self-Organized Criticality in a Deterministic Mechanical Model. *Physical Review A* 46 (1992) 6288.

Chapter 6

Alstrom, P., and Leao, J. Self-Organized Criticality in the Game of Life. *Physical Review E* 49 (1994) R2507.

Bak, P., Chen, K., and Creutz, M. Self-Organized Criticality in the Game of Life. *Nature* 342 (1989) 780.

Berlekamp, E., Conway, J., and Guy, R. *Winning Ways for Your Mathematical Plays,* vol. 2. New York: Academic, 1982.

Gardner, M. Mathematical Games. The Fantastical Combinations of John Conroy's New Solitaire Game "Life." *Scientific American* 223 (4) (1970) 120; (5) (1970) 114; 118 (6) (1970) 114.

Hemmingsen, J. Consistent Results on Life. *Physica D* 80, (1995) 80.

Chapter 7

Bak, P., Flyvbjerg, H., and Lautrup, B. Coevolution in a Rugged Fitness Landscape. *Physical Review A* 46 (1992) 6714.

Fisher, R. A. *The Genetical Theory of Natural Selection.* Oxford: Oxford University Press, 1932.

Kauffman, S. A. *The Origins of Order.* New York, Oxford: Oxford University Press, 1993.

Kauffman, S. A. and Johnsen, S. Coevolution to the Edge of Chaos—Coupled Fitness Landscapes, Poised States, and Coevolutionary Avalanches. *Journal of Theoretical Biology* 149 (1991) 467.

Smith, J. Maynard. *The Theory of Evolution.* Cambridge, Cambridge University Press, 1993.

Wright, S. Macroevolution—Shifting Balance Theory. *Evolution* 36 (1982) 427.

Chapter 8

Adami, C. Self-Organized Criticality in Living Systems. *Physical Letters A* 203 (1995) 29.

Alvarez, L. W., Alvarez, W., Asaro, F., and Michel, H. V. Extra Terrestrial Causes for the Cretaceous/Tertiary Extinctions. *Science* 208 (1980) 1095.

Alvarez, W. and Asaro, F. What Caused the Mass Extinction: an Extraterrestrial Impact. *Scientific American* 263 (1990) 76.

Bak, P., Flyvbjerg, H., and Sneppen, K. Can We Model Darwin? *New Scientist* 12 (1994) 36.

Bak, P. and Sneppen, K. Punctuated Equilibrium and Criticality in a Simple Model of Evolution. *Physical Review Letters* 24 (1993) 4083.

Collar, J. I. Biological Effects of Stellar Collapse Neutrinos. *Phys. Rev. Lett.* 76 (1996) 999.

Darwin, C. *The Origin of Species by Means of Natural Selection.* 6th ed. London: Appleton, 1910.

Kellogg, D. E. The Role of Phyletic Change in the Evolution of Pseudocubus— Vema Radiolaria. *Paleobiology* 1 (1975) 359.

Newman, M. E. J. and Roberts, B. W. Mass-Extinction: Evolution and the Effects of External Influences on Unfit Species. *Proceedings of the Royal Society B* 260 (1995) 31.

Paczuski, M., Maslov, S., and Bak, P. Avalanche Dynamics in Evolution, Growth, and Depinning Models. *Physical Review E* 53 (1995) 414.

Raup, D. M. and Sepkoski, J. J. Jr. Periodicity of Extinctions in the Geological Past. *Proceedings of the National Academy of Science, USA* 81 (1984) 801.

Sneppen, K., Bak, P., Flyvbjerg, H., and Jensen, M. H. Evolution as a Self-Organized Critical Phenomenon. *Proceedings of the National Academy of Science, USA* 92 (1995) 5209.

Vandewalle, N. and Ausloos, M. Self-Organized Criticality in Phylogenetic Tree Growths. *Journal de Physique I France* 5 (1995) 1011.

Chapter 9

Boettcher, S. and Paczuski, M. Exact Results for Spatio-Temporal Correlations in a Self-Organized Critical Model of Punctuated Equilibrium. *Physical Review Letters* 76 (1996) 348.

Deboer, J., Derrida, B., Flyvbjerg, H., Jackson, A., and Wettig, T. Simple Model of Self-Organized Biological Evolution. *Physical Review Letters* 73 (1994) 906.

Flyvbjerg, H., Sneppen, K., and Bak, P. Mean Field Theory for a Simple Model of Evolution. *Physical Review Letters* 71 (1993) 4087.

Ito, Keisuke. Punctuated Equilibrium Model of Biological Evolution is also a Self-Organized Critical Model of Earthquakes. *Physical Review E* 52 (1995) 3232.

Maslov, S., Paczuski, M., and Bak, P. Avalanches and $1/f$ Noise in Evolution and Growth Models. *Physical Review Letters* 73 (1994) 2162.

Paczuski, M., Maslov, S., and Bak, P. Field Theory for a Model of Self-Organized Criticality. *Europhysics Letters* 27 (1994) 97.

Chapter 10

Ashby, W. R. *Design for a Brain*, 2nd ed. New York: Wiley, 1960.

Stassinopoulos, D., and Bak, P. Democratic Reinforcement. A Principle for Brain Function. *Physical Review E* 51 (1995) 5033.

Chapter 11

Arthur, B. Increasing Returns, and Lock-ins by Historical Events. *The Economic Journal* 99 (1989) 116.

———. Positive Feedbacks in the Economy. *Scientific American* 262, February (1990) 92.

Axelrod, R. *The Evolution of Cooperation.* New York: Basic Books, 1984.

Bak, P., Chen, Kan, Scheinkman, J. A., and Woodford, M. Aggregate Fluctuations from Independent Shocks: Self-Organized Criticality in a Model of Production and Inventory Dynamics. *Ricerche Economiche* 47 (1993) 3.

Dhar, D. and Ramaswamy, R. Exactly Solved Model of Self-Organized Critical Phenomena. *Physical Review Letters* 63 (1989) 1659.

Nagel, K. and Paczuski, M. Emergent Traffic Jams. *Physical Review E* 51 (1995) 2909.

Scheinkman, J. A. and Woodford, M. Self-Organized Criticality and Economics Fluctuations. *American Journal of Economics* 84 (1994) 417.

Treiterer, J. *Aereal Traffic Photos.* Technical Report PB 246 094. Columbus, OH: Ohio State University, 1994.

Index